Electrical drives for direct drive renewable energy systems

Related titles:

Wind energy systems (ISBN 978-1-84569-580-4)

Stand-alone and hybrid wind energy systems (ISBN 978-1-84569-527-9)

Concentrating solar power technology (ISBN 978-1-84569-769-3)

Details of these books and a complete list of titles from Woodhead Publishing can be obtained by:

- visiting our web site at www.woodheadpublishing.com
- contacting Customer Services (e-mail: sales@woodheadpublishing.com; fax: +44 (0) 1223 832819; tel.: +44 (0) 1223 499140 ext. 130; address: Woodhead Publishing Limited, 80 High Street, Sawston, Cambridge CB22 3HJ, UK)
- contacting our US office (e-mail: usmarketing@woodheadpublishing.com; tel.: (215) 928 9112; address: Woodhead Publishing, 1518 Walnut Street, Suite 1100, Philadelphia, PA 19102–3406, USA)

If you would like e-versions of our content, please visit our online platform: www.woodheadpublishingonline.com. Please recommend it to your librarian so that everyone in your institution can benefit from the wealth of content on the site.

We are always happy to receive suggestions for new books from potential editors. To enquire about contributing to our Materials series, please send your name, contact address and details of the topic/s you are interested in to sarah.hughes@woodheadpublishing.com. We look forward to hearing from you.

The Woodhead team responsible for publishing this book:
Commissioning Editor: Sarah Hughes
Publications Coordinator: Lucy Beg
Project Editor: Sarah Lynch
Editorial and Production Manager: Mary Campbell
Production Editor: Adam Hooper
Freelance Project Manager: Bhavani Ganesh Kumar, Newgen Knowledge Works Pvt Ltd
Freelance Copyeditor: Nancy Richardson
Proofreader: Suma George, Newgen Knowledge Works Pvt Ltd
Cover Designer: Terry Callanan

Woodhead Publishing Series in Energy: Number 24

Electrical drives for direct drive renewable energy systems

Edited by
Markus Mueller and Henk Polinder

WOODHEAD PUBLISHING

Oxford Cambridge Philadelphia New Delhi

Published by Woodhead Publishing Limited,
80 High Street, Sawston, Cambridge CB22 3HJ, UK
www.woodheadpublishing.com
www.woodheadpublishingonline.com

Woodhead Publishing, 1518 Walnut Street, Suite 1100, Philadelphia,
PA 19102–3406, USA

Woodhead Publishing India Private Limited, G-2, Vardaan House, 7/28 Ansari Road,
Daryaganj, New Delhi – 110002, India
www.woodheadpublishingindia.com

First published 2013, Woodhead Publishing Limited

British Library Cataloguing in Publication Data
A catalogue record for this book is available from the British Library.

Library of Congress Control Number: 2013930286

ISBN 978-1-84569-783-9 (print)
ISBN 978-0-85709-749-1 (online)
ISSN 2044-9364 Woodhead Publishing Series in Energy (print)
ISSN 2044-9372 Woodhead Publishing Series in Energy (online)

The publisher's policy is to use permanent paper from mills that operate a sustainable forestry policy, and which has been manufactured from pulp which is processed using acid-free and elemental chlorine-free practices. Furthermore, the publisher ensures that the text paper and cover board used have met acceptable environmental accreditation standards.

Typeset by Newgen Knowledge Works Pvt Ltd
Printed by MPG Printgroup, UK

Contents

Contributor contact details

(* = main contact)

Editors

Professor Markus Mueller
Institute for Energy Systems
School of Engineering
Mayfield Road
University of Edinburgh
Edinburgh
EH9 3JL
UK

E-mail: markus.mueller@ed.ac.uk

Dr Henk Polinder
Electrical Power Processing
Electrical Engineering, Mathematics and Computer Science
Delft University of Technology
Mekelweg 4
2628 CD Delft
The Netherlands

E-mail: H.Polinder@tudelft.nl

Chapter 1

Professor Markus Mueller* and Aristides Zavvos
Institute for Energy Systems
School of Engineering
Mayfield Road
University of Edinburgh
Edinburgh
EH9 3JL
UK

E-mail: markus.mueller@ed.ac.uk

Chapter 2

Dr Henk Polinder
Electrical Power Processing
Electrical Engineering, Mathematics and Computer Science
Delft University of Technology
Mekelweg 4
2628 CD Delft
The Netherlands

E-mail: H.Polinder@tudelft.nl

Chapter 3

Dr Alasdair McDonald*
Wind Energy Systems Doctoral Training Centre
Department for Electronic and Electrical Engineering
Room 3-36, Royal College Building
University of Strathclyde
204 George Street
Glasgow
G1 1XW
UK

E-mail: alasdair.mcdonald@strath.ac.uk

Professor Markus Mueller and Aristides Zavvos
Institute for Energy Systems
School of Engineering
Mayfield Road
University of Edinburgh
Edinburgh
EH9 3JL
UK

E-mail: markus.mueller@ed.ac.uk

Chapters 4 and 5

Dr Z. Chen
Department of Energy Technology
Aalborg University
Pontoppidanstraede 101
DK-9220 Aalborg East
Denmark

E-mail: zch@et.aau.dk

Chapter 6

E. de Vries
Rotation Consultancy
Van Lenneplaan 8
3818VE Amersfoort
The Netherlands

E-mail: rotation.edv@planet.nl

Chapter 7

Ir. Anoop Jassal*
LB.03.670
EWI Faculty
TU Delft
Mekelweg 4
2628CD
Delft
The Netherlands

E-mail: anoop.jas@gmail.com

Dr Henk Polinder
Electrical Power Processing
Electrical Engineering, Mathematics and Computer Science
Delft University of Technology
Mekelweg 4
2628 CD Delft
The Netherlands

E-mail: H.Polinder@tudelft.nl

Kees Versteegh
XEMC-Darwind
The Netherlands

E-mail: k.versteegh@xemc-darwind.com

Chapters 8 and 9

Miguel Prado* and Dr Henk Polinder
Electrical Power Processing
Electrical Engineering, Mathematics and Computer Science
Delft University of Technology
Mekelweg 4
2628 CD Delft
The Netherlands

E-mail: miguel.prado@teamwork.nl; H.Polinder@tudelft.nl

Chapter 10

Ozan Keysan
Institute for Energy Systems
University of Edinburgh
Mayfield Road, King's Buildings
Edinburgh
EH9 3JL
UK

E-mail: o.keysan@ed.ac.uk

Woodhead Publishing Series in Energy

1 **Generating power at high efficiency: Combined cycle technology for sustainable energy production**
Eric Jeffs

2 **Advanced separation techniques for nuclear fuel reprocessing and radioactive waste treatment**
Edited by Kenneth L. Nash and Gregg J. Lumetta

3 **Bioalcohol production: Biochemical conversion of lignocellulosic biomass**
Edited by K. W. Waldron

4 **Understanding and mitigating ageing in nuclear power plants: Materials and operational aspects of plant life management (PLiM)**
Edited by Philip G. Tipping

5 **Advanced power plant materials, design and technology**
Edited by Dermot Roddy

6 **Stand-alone and hybrid wind energy systems: Technology, energy storage and applications**
Edited by J. K. Kaldellis

7 **Biodiesel science and technology: From soil to oil**
Jan C. J. Bart, Natale Palmeri and Stefano Cavallaro

8 **Developments and innovation in carbon dioxide (CO_2) capture and storage technology Volume 1: Carbon dioxide (CO_2) capture, transport and industrial applications**
Edited by M. Mercedes Maroto-Valer

9 **Geological repository systems for safe disposal of spent nuclear fuels and radioactive waste**
Edited by Joonhong Ahn and Michael J. Apted

10 **Wind energy systems: Optimising design and construction for safe and reliable operation**
Edited by John D. Sørensen and Jens N. Sørensen

11 **Solid oxide fuel cell technology: Principles, performance and operations**
Kevin Huang and John Bannister Goodenough

12 **Handbook of advanced radioactive waste conditioning technologies**
Edited by Michael I. Ojovan

13 **Membranes for clean and renewable power applications**
Edited by Annarosa Gugliuzza and Angelo Basile
14 **Materials for energy efficiency and thermal comfort in buildings**
Edited by Matthew R. Hall
15 **Handbook of biofuels production: Processes and technologies**
Edited by Rafael Luque, Juan Campelo and James Clark
16 **Developments and innovation in carbon dioxide (CO_2) capture and storage technology Volume 2: Carbon dioxide (CO_2) storage and utilisation**
Edited by M. Mercedes Maroto-Valer
17 **Oxy-fuel combustion for power generation and carbon dioxide (CO_2) capture**
Edited by Ligang Zheng
18 **Small and micro combined heat and power (CHP) systems: Advanced design, performance, materials and applications**
Edited by Robert Beith
19 **Advances in clean hydrocarbon fuel processing: Science and technology**
Edited by M. Rashid Khan
20 **Modern gas turbine systems: High efficiency, low emission, fuel flexible power generation**
Edited by Peter Jansohn
21 **Concentrating solar power technology: Principles, developments and applications**
Edited by Keith Lovegrove and Wes Stein
22 **Nuclear corrosion science and engineering**
Edited by Damien Féron
23 **Power plant life management and performance improvement**
Edited by John E. Oakey
24 **Electrical drives for direct drive renewable energy systems**
Edited by Markus Mueller and Henk Polinder
25 **Advanced membrane science and technology for sustainable energy and environmental applications**
Edited by Angelo Basile and Suzana Pereira Nunes
26 **Irradiation embrittlement of reactor pressure vessels (RPVs) in nuclear power plants**
Edited by Naoki Soneda
27 **High temperature superconductors (HTS) for energy applications**
Edited by Ziad Melhem

28 **Infrastructure and methodologies for the justification of nuclear power programmes**
Edited by Agustín Alonso

29 **Waste to energy (WtE) conversion technology**
Edited by Marco Castaldi

30 **Polymer electrolyte membrane and direct methanol fuel cell technology Volume 1: Fundamentals and performance of low temperature fuel cells**
Edited by Christoph Hartnig and Christina Roth

31 **Polymer electrolyte membrane and direct methanol fuel cell technology Volume 2: *In situ* characterization techniques for low temperature fuel cells**
Edited by Christoph Hartnig and Christina Roth

32 **Combined cycle systems for near-zero emission power generation**
Edited by Ashok D. Rao

33 **Modern earth buildings: Materials, engineering, construction and applications**
Edited by Matthew R. Hall, Rick Lindsay and Meror Krayenhoff

34 **Metropolitan sustainability: Understanding and improving the urban environment**
Edited by Frank Zeman

35 **Functional materials for sustainable energy applications**
Edited by John A. Kilner, Stephen J. Skinner, Stuart J. C. Irvine and Peter P. Edwards

36 **Nuclear decommissioning: Planning, execution and international experience**
Edited by Michele Laraia

37 **Nuclear fuel cycle science and engineering**
Edited by Ian Crossland

38 **Electricity transmission, distribution and storage systems**
Edited by Ziad Melhem

39 **Advances in biodiesel production: Processes and technologies**
Edited by Rafael Luque and Juan A. Melero

40 **Biomass combustion science, technology and engineering**
Edited by Lasse Rosendahl

41 **Ultra-supercritical coal power plant: Materials, technologies and optimisation**
Edited by Dongke Zhang

42 **Radionuclide behaviour in the natural environment: Science, implications and lessons for the nuclear industry**
Edited by Christophe Poinssot and Horst Geckeis

43 **Calcium and chemical looping technology for power generation and carbon dioxide (CO_2) capture: Solid oxygen- and CO_2-carriers**
P. Fennell and E. J. Anthony

44 **Materials' ageing and degradation in light water reactors: Mechanisms, and management**
Edited by K. L. Murty

45 **Structural alloys for power plants: Operational challenges and high-temperature materials**
Edited by Amir Shirzadi, Rob Wallach and Susan Jackson

46 **Biolubricants: Science and technology**
Jan C. J. Bart, Emanuele Gucciardi and Stefano Cavallaro

47 **Wind turbine blade design and materials: Improving reliability, cost and performance**
Edited by Povl Brøndsted and Rogier Nijssen

48 **Radioactive waste management and contaminated site clean-up: Processes, technologies and international experience**
Edited by William E. Lee, Michael I. Ojovan, Carol M. Jantzen

49 **Probabilistic safety assessment for optimum nuclear power plant life management (PLiM): Theory and application of reliability analysis methods for major power plant components**
Gennadij V. Arkadov, Alexander F. Getman and Andrei N. Rodionov

50 **The Coal handbook Volume 1: Towards cleaner production**
Edited by D. G. Osborne

51 **The Coal handbook Volume 2: Coal utilisation**
Edited by D. G. Osborne

52 **The biogas handbook: Science, production and applications**
Edited by Arthur Wellingerr, Jerry Murphy and David Baxter

53 **Advances in biorefineries: Biomass and waste supply chain exploitation**
Edited by K. W. Waldron

54 **Geoscience of carbon dioxide (CO_2) storage**
Edited by Jon Gluyas and Simon Mathias

55 **Handbook of membrane reactors Volume 1: Fundamental materials science, design and optimisation**
Edited by Angelo Basile

56 **Handbook of membrane reactors Volume 2: Reactor types and industrial applications**
Edited by Angelo Basile

57 **Alternative fuels and advanced vehicle technologies: Towards zero carbon transportation**
Edited by Richard Folkson

58 **Handbook of microalgal bioprocess engineering**
Christopher Lan and Bei Wang

59 **Fluidized bed technologies for near-zero emission combustion and gasification**
Edited by Fabrizio Scala

60 **Managing nuclear projects: A comprehensive management resource**
Edited by Jas Devgun

61 **Handbook of process integration: Minimisation of energy and water use, waste and emissions**
Edited by Jiří Klemeš

Part I
Electrical drive technology

1
Electrical generators for direct drive systems: a technology overview

M. MUELLER and A. ZAVVOS,
University of Edinburgh, UK

DOI: 10.1533/9780857097491.1.3

Abstract: This chapter gives an overview of the various direct drive generator technologies for wind turbines and focuses on the different permanent magnet generator topologies. Their advanced characteristics have drawn a number of manufacturers towards this system which now represents 20% of the sold wind turbines worldwide. However, in order to attain high torque levels they require an increased airgap diameter. This chapter will guide the reader through the various direct drive generator topologies for wind turbines that have been suggested in the literature and outline the ones with the highest potential to produce large power output with the least possible weight.

Key words: direct drive, permanent magnet generator, transverse flux machines, switched reluctance generator, radial flux, axial flux.

1.1 Introduction

In recent years, large-scale direct drive generators have attracted the attention of wind turbine manufacturers as an alternative to geared systems. In a direct drive system the gearbox is removed from the drive train and the generator is directly coupled to the hub of the wind turbine (Fig. 1.1). They both rotate at the same speed which typically varies between 8 and 15 rpm depending upon the power rating.

Elimination of the gearbox has several benefits, including reduced noise levels, simplification of the drive train, which increases reliability, reduced losses due to fewer energy conversion steps, and lower maintenance cost.[1,2] Costly gearbox maintenance issues that can cause long downtime periods, such as oil replacement, gearbox failures or gearbox replacement, are avoided.[3–5] Furthermore, the reduction in the number of bearings and moving parts required for direct drive systems results in Reference 6. These advanced characteristics have attracted a number of manufacturers to this system, which now represents 20% of the wind turbines sold worldwide.[7]

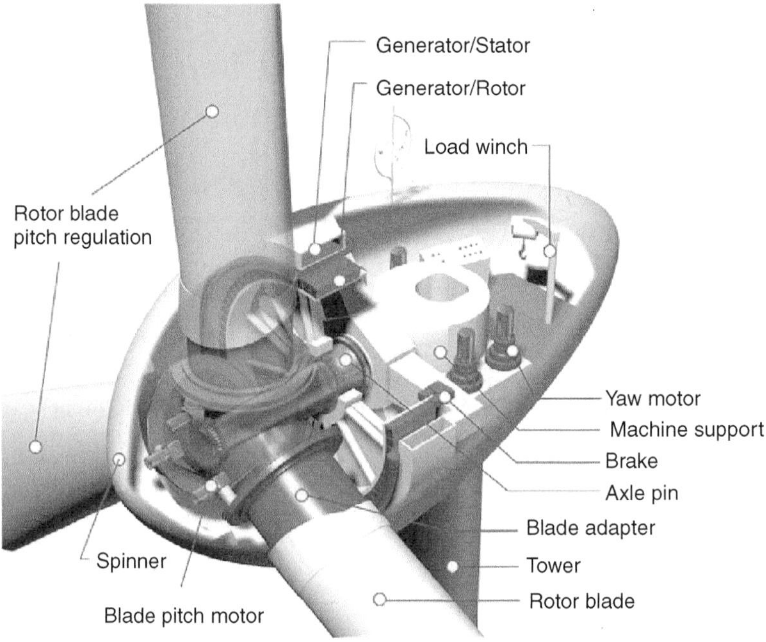

1.1 The direct drive system.[4]

In order to compete with high-speed geared generators, direct drive machines need to attain high torque levels. Equation [1.1] gives the power output of a rotating machine.

$$P = \omega T \qquad [1.1]$$

where P, the power output (W), ω, the angular velocity (rad/s) and T, the rated torque of the machine (Nm). When ω is low, T needs to increase to create the same power output. Equation [1.2] gives the torque of a rotating machine.

$$T = 2\pi R^2 l F_{\mathrm{d}} \qquad [1.2]$$

where R, the radius of the machine (m), l, the axial length (m) and F_{d} (Pa), the shear stress applied on the machine's structure due to the interaction of the machine's magnetic field and the currents generated (Fig. 1.2).

The combination of Equations [1.1] and [1.2] leads to the conclusion that since direct drive machines operate at a low rotational speed, given a fixed axial length and shear stress, it is necessary to increase the machine's diameter in order to achieve the required torque levels. The large diameters required to achieve high power outputs in direct drive machines need large amounts of expensive raw materials. Their very large size also makes them heavy and difficult to build, transport and install.

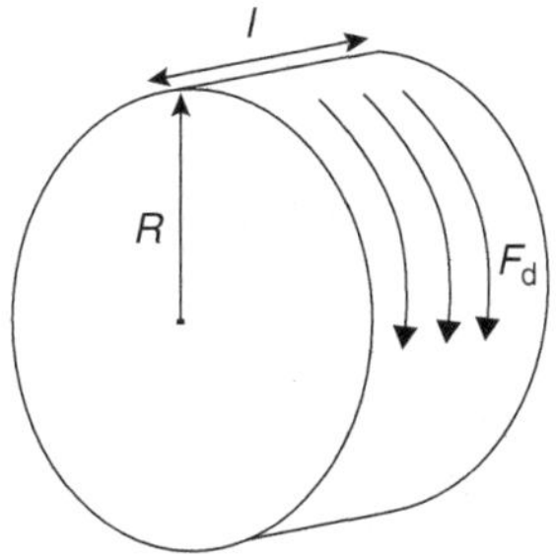

1.2 The dimensions of a rotating machine and the acting shear stress.

Although the structural loads applied to these machines do not differ from those of conventional wind turbine generators, it is their size that increases the magnitude of these forces and makes them so structurally demanding. The structure has to be stiff and robust in order to maintain the small airgap clearance between the rotor and the stator against the various structural forces and at the same time hold the electromagnetically active materials in place against the attraction forces and gravity.[8] McDonald showed that the structural mass for a direct drive generator with nominal output of 5 MW can reach up to 80% of its total weight.[9]

Different direct drive configurations have been suggested in the literature in an attempt to produce a highly efficient machine with reduced structural mass. The aim of this chapter is to present the proposed direct drive generator topologies for wind turbines and describe those with the highest potential to produce large power outputs at the least possible weight.

1.2 Excitation methods

An AC synchronous machine can be electrically excited or permanent magnet excited. A switched reluctance generator (SRG) has single electrical excitation on the stator without any excitation on its rotor.

1.2.1 Electrically excited direct drive (EEDD) generators

Magnetization of the rotor poles of an electrically excited direct drive (EEDD) generator is provided by a DC source. DC excitation is usually provided via slip rings and brushes. Brushless DC excitation is also possible by employing a rotary AC exciter connected to the rotor through a bridge rectifier, but this arrangement is less common for wind turbine generators.[10]

The rotor poles of an EEDD machine can be salient or cylindrical, the former being most commonly used. In all cases, the rotor poles of the EEDD generator must be large enough to provide adequate space for the excitation windings. The stator of an EEDD is similar to that of an induction generator with three-phase distributed winding inserted into a slotted laminated iron core.[11]

A power converter is used to process the generated power and connect the EEDD generator to the grid. The amplitude and frequency of the voltage, as well as the active and reactive power of the machine, can be fully controlled even at fault grid conditions. Furthermore, the generator speed can be fully controllable for a wide range of wind speeds.[12,13]

Electrically excited synchronous machines are robust and simple to construct. Furthermore, for large power outputs, EEDD machines have a better power factor and efficiency compared to an induction machine. However, the need for constant direct current supply to the main field winding leads to additional losses of generated heat, reducing overall efficiency.[11] On the other hand, the external electrical excitation can be adjusted according to the prevailing wind conditions, keeping the voltage at rated values for low or high wind speeds.

The EEDD technology has matured over the last decade and is now the dominant technology for low-speed direct drive applications in the wind turbine market.[11] Enercon is the main supplier of this generator type, accounting for more than 15% of the total market and 75% of direct drive applications currently installed.[7] Enercon's prototype direct drive ystem E-126[14] (Fig. 1.3), with a blade diameter of 127 m and generator diameter in the order of 12 m, can reach a power output up to 7.5 MW. MTorres also produces EEDD systems with power outputs up to 2.5 MW.[16]

1.2.2 Permanent magnet direct drive (PMDD) generators

Electrical excitation of the rotor poles of a direct drive machine, brushless or not, can lead to resistive heat losses in the system. These losses reduce

1.3 Enercon E-126 7MW WEC.[15]

the system's efficiency and may cause maintenance issues. To avoid resistive heating in the direct drive system and the complicated cooling schemes required, a number of manufacturers have turned to systems with permanent magnets.[17]

In a PMDD machine, the rotor poles are made of permanent magnet material, therefore no external power supply is needed. This eliminates excitation losses in the generator and decreases the heat developed in the system. The energy yield and the overall efficiency are thus increased while the absence of slip rings increases the reliability of the machine. Smaller pole pitches can also be used, reducing the size of the generator.[18] On the downside, PM materials are expensive and difficult to handle in manufacture.[4] The stator of a PMDD is usually identical to that of an EEDD generator, but alternative stator topologies have been proposed as well.[19,20] A full-scale power converter is required for their connection to the grid. Advances in power electronics have decreased the cost of such power converters and allow PMDD generators to produce a clean power output.[21,22]

The rotor poles of a PMDD machine are made of rare earth materials such as samarium cobalt (SmCo) or neodymium iron boron (NdFeB) that exhibit high magnetic energy densities within a small volume and geometry. SmCo magnets are mainly used in high-temperature applications. Vilsboll *et al.* concluded in favour of the NdFeB magnets as they produce a greater remanent flux density – 1.2 T instead of 1.0 T – and can reduce the overall mass and price of the PMDD generator even further.[23]

The high cost of PM materials originally prevented many manufacturers from employing this type of excitation for their machines. The price of magnets fell between 1995 and 2005 almost by a factor of 10, increasing their use in a number of commercial or military applications.[24] However, 95% of rare earths used in such magnets are mined in China. The current monopoly that China enjoys, together with today's increased demand, has raised prices once more and created a generalized uncertainty regarding their extended use. Nevertheless, the future of PMDD generators seems to be promising, as rare earths are now being found at an increasing number of sites.[25]

Permanent magnet excitation produces a robust and simple machine with superior efficiency and torque density, and reduced whole life cost. The drop in prices of PM materials and power converters has encouraged a number of developers to turn to this excitation type for direct drive machines over the last decade. GE recently acquired Scanwind in order to expand into the PMDD generator market.[25] The proposed 4.0 MW design was specifically for offshore purposes (Fig. 1.4).[26] Siemens (3.0 MW) have also turned to PM excitation for their direct drive wind turbine systems.[27] Goldwind (1.5–2.5 MW), STX Windpower (1.5–2.0 MW), Emergya Wind Technologies

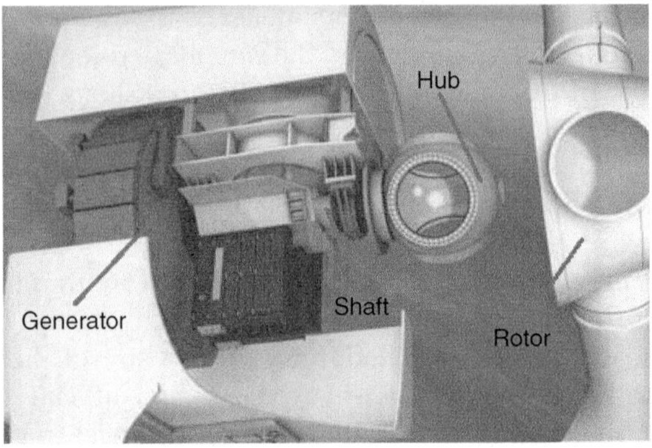

1.4 The interior parts of the 4 MW GE (former Scanwind) wind turbine.[26]

(0.5–2.0 MW), Vensys (1.5–2.5 MW), Leitwind (1.5–3.0 MW) and XEMC Darwind (5.0 MW) are some of the wind turbine companies that manufacture PMDD generators in the MW scale.[28–33]

Due to the rapid growth of commercial and military interest in PM technology over the last decade, the industrial base for high-power PM machines has also increased, and this is underlined by the increasing number of PMDD manufacturers who have established the PM synchronous generator as a prime candidate for direct drive wind turbine applications.[10]

1.2.3 Switched reluctance direct drive generators

The operating principle of the SRG is based on the tendency of a magnetic field to follow the path of least reluctance. In an SRG, the rotor of the machine 'tries' to align with the stator poles to obtain the lowest magnetic reluctance (Fig. 1.5).[10,35] The stator and rotor of the SRG are laminated to minimize eddy currents and of salient construction, which provides the variations in airgap length around the circumference. It is this saliency that causes rotation by the rotor's attempt to create the lowest reluctance path for the flux to travel.[10] The SRG is an inherently variable speed machine in which the output can be controlled by the switching instants, whether operating as a generator or motor.

There are no coils on the rotor. Concentrated coils are wound around each stator pole, and excited with DC. SRG has been considered for direct drive wind turbine generation because it is simple to manufacture, highly robust, easy to cool and has a cheap design.[36–38] On the downside, it has a lower torque density compared to permanent magnet machines; it requires excitation from the grid; the machine exhibits torque ripple and noise, both

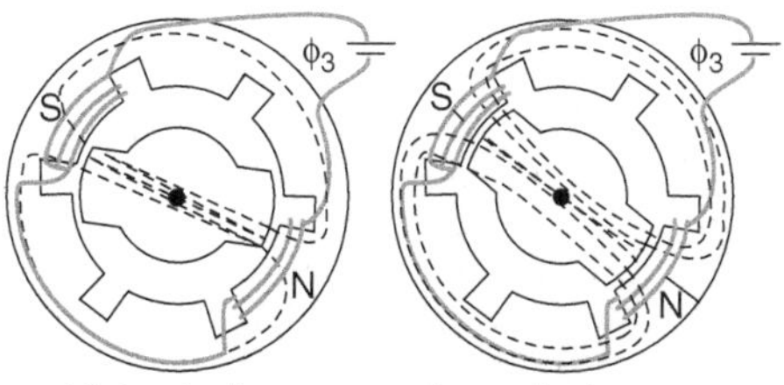

1.5 The operating principle of the SRG.[34]

of which can be overcome by suitable control. The low torque density compared to permanent magnet machines has prevented manufacturers from considering the reluctance machine as an option for direct drive power take-off systems for wind turbine applications.[10]

1.3 Permanent magnet direct drive (PMDD) generator topologies

PMDD generators seem to have the greatest potential for onshore and offshore direct drive wind development, compared to EEDD and SRG machines, due to their advanced characteristics. Nevertheless, their large size and heavy structures remain fundamental issues. A great many different PMDD topologies have been proposed in the literature in an attempt to produce machines with high torque/mass or power/cost ratio. The easiest way to categorize them is by the orientation of the magnetic flux as it crosses the airgap within the machine, leading to radial flux (RF), axial flux (AF) and transverse flux (TF) topologies. Depending on the stator's core design, the machine can be slotted or slotless.[20] Another way to categorize PMDD generators is by the presence or absence of iron in the stator's core, resulting in iron cored or air cored machines accordingly.[39,40]

1.3.1 Radial flux (RF) PMDD machines

In an RF machine magnetic flux flows radially across the airgap. The iron cored RF machine, with surface mounted permanent magnet poles rotating inside stationary armature windings, is the most common topology for PMDD generators because of its structural stability and robust design (Fig. 1.6). The slotted RF PMDD machine is the most conventional one, as it combines the structural characteristics of an EEDD machine with the advanced magnetic characteristics of the permanent magnets. The reduced weight for high torque ratings has established RF machines as the most common option for industrial PMDD generators for wind applications on the MW scale.[5,7,10,17,19]

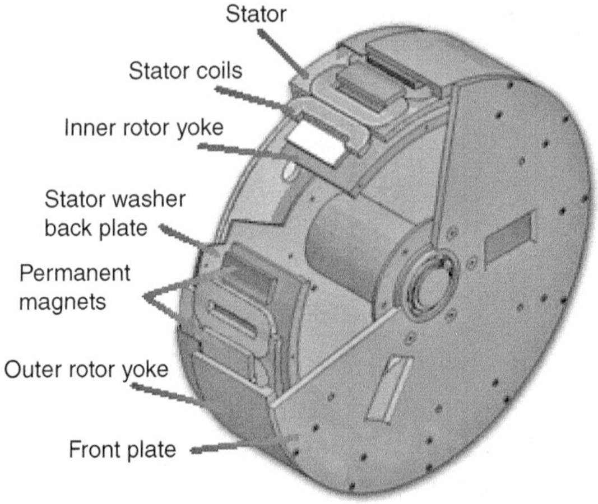

1.6 The RF PMDD machine.

However, alternative topologies have been suggested with non-conventional permanent magnet placement, rotor position or stator core design.

Spooner *et al.* suggested flux concentration methods to reduce the active material requirement and the total cost of RF PMDD machines.[41–43] The use of flux concentrators allows a higher flux density in the airgap than the remanent flux density of the permanent magnets. Steel concentrators also house the permanent magnets on the rotor and transfer the torque and thermal energy to the support structure (Fig. 1.7). This topology allows ferrite magnets to be used, which are much cheaper than NdFeB magnets.[44,45] However, burying the magnets together with the flux concentrators is difficult to achieve and creates a complex structure with manufacturing issues.

The rotor and the stator of an RF PMDD can be alternatively positioned with the rotor on the outer side of the machine (Fig. 1.8). Outer rotor RF PMDD generators permit a larger airgap diameter for the same dimensions of a conventional inner rotor RF PMDD machine.[46–49] This improves the machine's efficiency as it allows a larger number of magnetic poles to be used. On the downside, they require stiffer structures, which increases the total mass and complexity. Furthermore, as there is no natural cooling of the stator, complex cooling systems such as liquid cooling are necessary, which introduces additional reliability issues and maintenance cost.[4]

In a conventional machine, the copper windings of the stator are placed in the stator's vertical slots. In a slotless machine, however, the windings have a toroidal shape and are placed in flat recesses. A slotless stator topology with a double rotor was suggested by Korouji.[50] This topology is characterized by short end windings that reduce the overall weight and cost of the active material. The absence of teeth reduces the iron losses in the machine

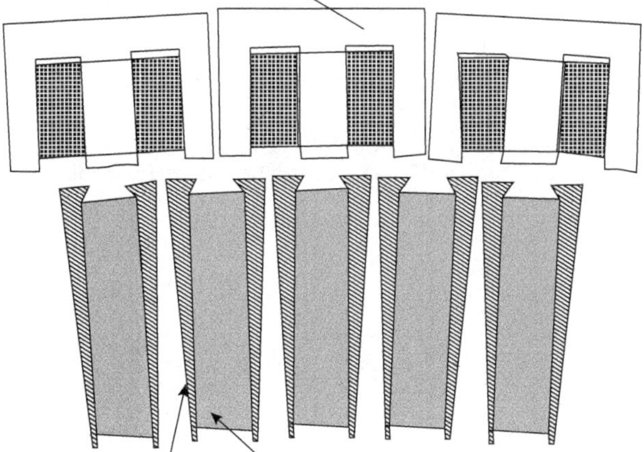

1.7 The RF PMDD machine with flux concentrators housing the PM material.[42,43]

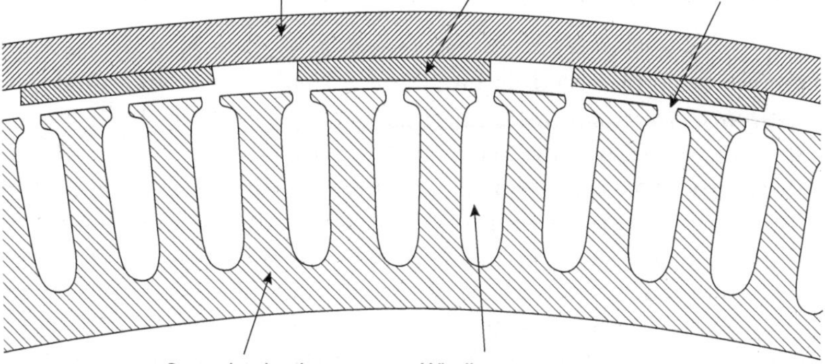

1.8 Outer rotor RF PMDD machine.[49]

and increases the overall efficiency,[51] although it also increases eddy current losses in the coil.

Another suggested way to reduce the cost of RF PMDD generators is to use fractional pitch windings as illustrated in Fig. 1.9.[52] This topology is expected to reduce cost due to the lower number of simpler coils around the stator's teeth compared to conventional complicated end windings. On the downside, more eddy current losses in the magnets and the back iron are generated in this machine type due to increased sub harmonics.

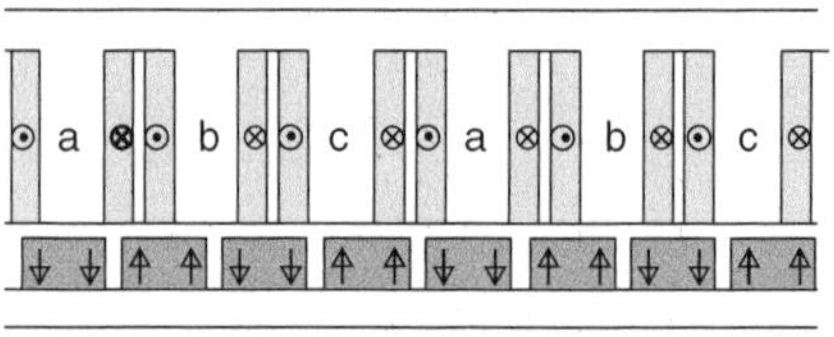

1.9 Sketch of a cross-section of eight poles of a permanent magnet synchronous machine with a cheap fractional pitch winding.[4]

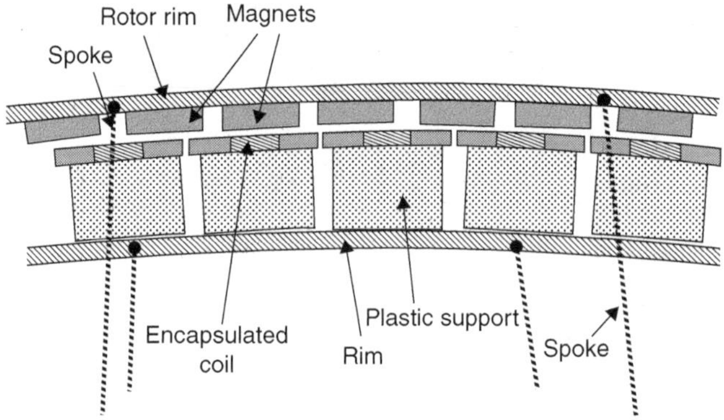

1.10 Cross-section of suggested ironless RF PMDD machine.[53]

Spooner *et al.* proposed an outer rotor topology for large direct drive wind turbines, with an ironless stator and a pair of spoked wheels that carry the structure.[53] The machine's cross-section is depicted in Fig. 1.10. The elimination of iron from the stator releases the structure from the large attractive forces and effectively reduces the required stiffness reducing the mass to that of a geared generator. Manufacture is also simplified and cogging torque is reduced to zero. On the downside, without a stator core the magnetic flux cannot easily cross the airgap, thus larger volumes of permanent magnet are required. Eddy current losses in the windings are also likely to be higher for such a machine. Furthermore, the suggested spokes can be aerodynamically inefficient for large diameters.

Mechanical support that takes the loading path from the generator structure and reduces the required stiffness of the machine has been reported in the literature for RF PMDD machines.[54–58] In these concepts, bearings are placed near the airgap, increasing the load capacity of the structure and contributing to airgap management. Thus the structural weight required for the rotor and stator can be reduced. Versteegh suggested using large diameter bearings near the airgap for the STX 72 machine (formerly Harakosan Z72).[54] Although these bearings would reduce the total weight of the direct

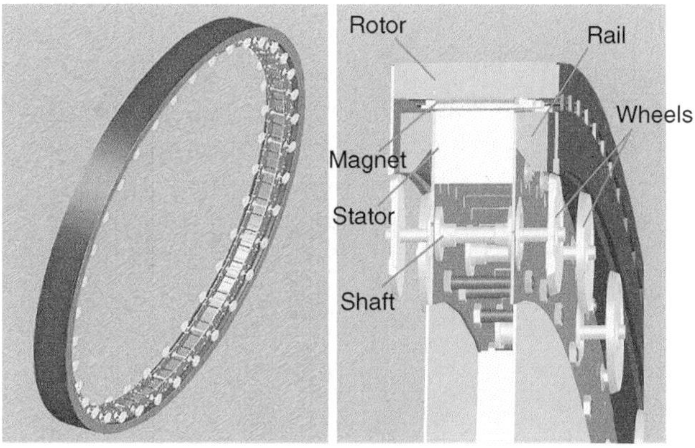

1.11 The NewGen generator concept.[55]

drive generator, the required stiffness for large diameters would make them heavy and expensive.[56] Engstrom and Lindgren suggested wheels at the airgap that are supported on a rail which is embedded in the stator structure (Fig. 1.11).[55] The NewGen design resulted in a significant reduction of the mass and the stiffness of the direct drive generator. However, the large number of wheels running at high speeds would raise the temperature of the bearings and create maintenance and reliability issues due to vibration.

Shrestha *et al.* suggested replacing the bearing wheels with non-contact magnetic bearings that levitate the machine, eliminating the unreliable gear wheels at the airgap (Fig. 1.12).[57] Torque carriers and a mechanical bearing are also employed. This assembly is expected to reduce the structural mass of a 5 MW RF PMDD generator by more than 40%. Since magnetic bearings require active control, this concept would introduce additional control issues and increase the total cost of the direct drive machine due to the expensive magnetic bearings.[58]

Despite the novelty of all the alternative concepts, a number of researchers who have compared the various RF designs conclude in favour of the conventional RF PMDD machine with permanent magnets mounted on the surface of the rotor because of its structural simplicity; the high energy yield torque density and reliability; and the reduced manufacture or maintenance costs.[6,10,16,19,44,58]

1.3.2 Axial flux (AF) PMDD machines

In an AF machine the magnetic flux flows in the axial direction across the airgap. Disc structures are most commonly employed. The slotted AF PMDD machine with the PM material mounted on the rotor and the stator facing

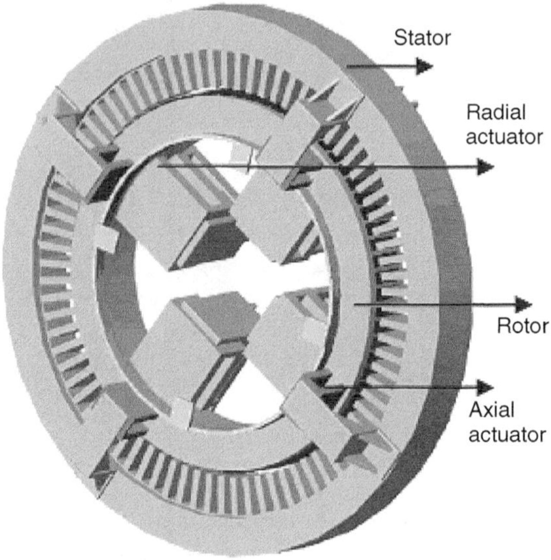

1.12 View of the magnetic bearing setup for an iron cored RF PMDD generator.[56]

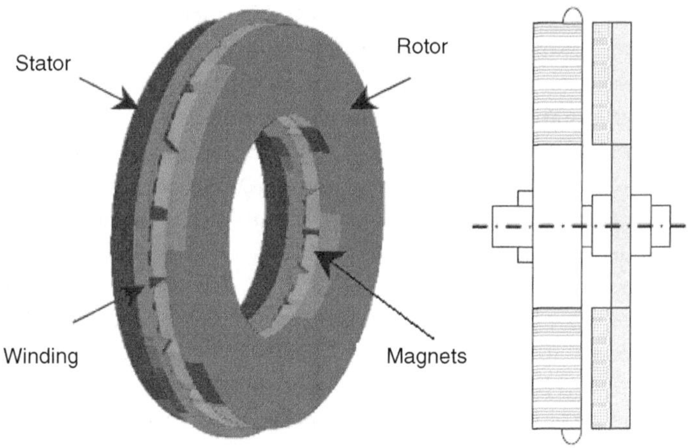

1.13 Single-sided slotted AF PMDD machine.[71]

the rotor is the basic structural design for this generator type (Fig. 1.13). Their characteristics and technological evolution from the early 1980s to the present are described in References 59–64.

Slotted AF PMDD machines offer a compact design with relatively low cogging torque and noise, short axial length and a high torque density.[65–69] In their extended comparison of different PMDD generator technologies,

1.14 Single-sided slotted AF PMDD machine with stator balance.[70]

Dubois *et al.* conclude that slotted AF PMDD machines have a lower cost/torque ratio compared to conventional RF machines.[44]

A fundamental issue with single-sided AF machines is the large magnetic attraction force between the PM disc and the iron stator which increases the stiffness requirements for this machine type. An additional rotor or stator is commonly used to form a double-sided machine in which the forces are balanced, preventing the rotor from moving towards the stator and vice versa (Fig. 1.14). This produces machines that are heavier than their RF counterparts.[70]

Alternative double-sided topologies with a double stator or a double rotor structure have been proposed in the literature for slotted AF PMDD machines. Bang made a structural comparison between slotted AF designs for large power outputs and concluded that the double-stator design with permanent magnets mounted on both sides of the rotor seems to be a lighter slotted topology compared to single-sided or double-rotor topologies (Fig. 1.15).[19]

However, the complicated core designs make slotted AF machines difficult to manufacture compared to RF PMDD machines. In addition, a second magnetic flux path that moves along with the main AF and is both radially and axially oriented is created in such topologies due to magnetic flux leakage.[71] The resulting complex three-dimensional electromagnetic design is hard to calculate and makes it difficult to accurately predict the performance of such machines. These attributes, in addition to the difficulty of maintaining the airgap in large diameters and the complicated cooling systems that are required, make slotted AF PMDD machines even less favourable.[10] An efficient way to minimize these characteristics is the removal of the stator's teeth from the structure.

1.15 Double-sided slotted AF PMDD machine with double stator.[70]

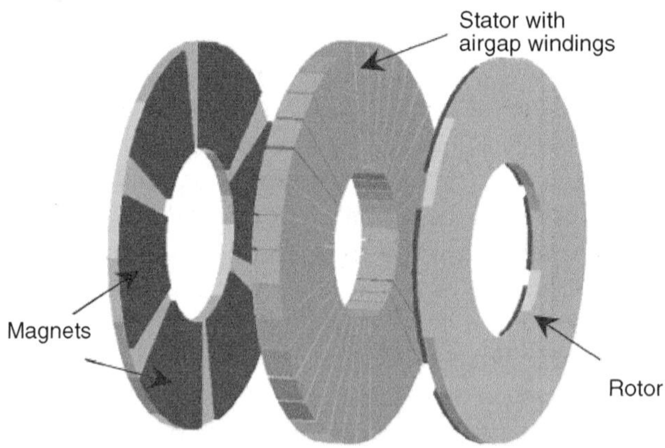

1.16 The double-sided AF PMDD TORUS machine.[71]

The slotless AF PMDD generator with toroidal windings (TORUS) is described by Spooner and Chalmers.[72,73] The stator of TORUS consists of a stack of laminated steel with the windings wrapped around its core in a toroidal manner. A double-sided structure with a single stator between two rotors with permanent magnets mounted on them is the preferred topology for such a machine (Fig. 1.16).

The slotless stator design offers short end windings, which reduces copper losses, increases the overall efficiency and creates a more compact design with a shorter axial length that is easier and cheaper to manufacture.[70–72] Effects caused by the presence of the slots, such as cogging torque, flux ripple and high frequency rotor loss, are eliminated. In addition, the TORUS concept has twice the torque/mass ratio of a conventional RF

PMDD machine due to the reduced structural mass. However, this design has a low power density, making necessary the use of large outer diameters and thicker magnets for compensation.[70] TORUS has therefore a high cost/torque ratio which can be twice as large as that of a common RF machine for any given diameter.[44] Thus this configuration would not be cost effective for high power rated wind turbines with large diameters.

Ironless – or 'air cored' as they are more commonly called – AF PMDD machines have also been suggested for wind applications.[39] The double-sided topology with two rotors and a stator is also favoured for this generator design. The stator structure is eliminated in this case and the conducting material is embedded in a pocket made of non-magnetic material such as glass fiber reinforced epoxy.[58] An ordinary rotor is employed with the permanent magnets mounted on its disc. The rotor and stator structures are relieved from maintaining the airgap clearance against strong attraction forces in such configuration, leading to a more compact and lightweight design that is easy to manufacture and assembly. The efficiency of the generator is also increased since the cogging torque and possible iron losses are reduced to zero.[76] An air cored arrangement leads to very low machine inductances and negligible armature reaction compared to TORUS machines.[77] On the downside, without an iron core to aid the movement of the flux across the airgap, larger outer diameters and greater volumes of permanent magnets are required to establish the necessary flux density in the airgap.[19]

TORUS and aircored machines can be constructed with multiple stages for applications with small diameter and high torque requirements. Multistage AF PMDD topologies are reviewed in References 69 and 76–80. To create 'n' stages in a machine, n stators are placed between $n + 1$ rotors that are mounted on the same shaft.

A three-stage TORUS design is illustrated in Fig. 1.17. However multistage configurations would have to compensate for the large attraction forces in the case of an imbalance in the airgaps on either side of the stator. Even for air cored multistage machines, there will be a force imbalance between the outer rotors and the stators. In addition, the small flux density in the airgap should be significantly improved for large power output applications and the structure should have the adequate stiffness to withstand possible unbalanced load sharing among the stages.[83]

Mueller *et al.* suggested in Reference 8 a C-core design for an air cored AF PMDD machine with the C-cores mounted on the rotor structure while the stator winding is held independently between them (Fig. 1.18). The C-core orientation could be radially or axially oriented. The AF orientation has been chosen as it offers extra structural simplicity and can be more easily extended to multistage machines.

A significant mass reduction of up to 55% on the inactive mass of the generator is suggested for this concept without compromising the structure's

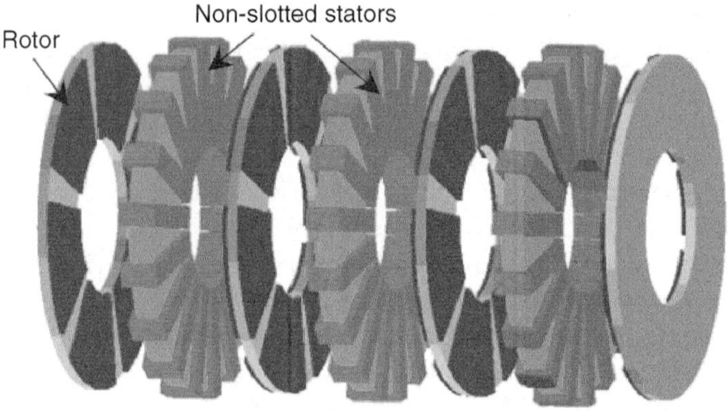

1.17 A three-stage TORUS machine.[71]

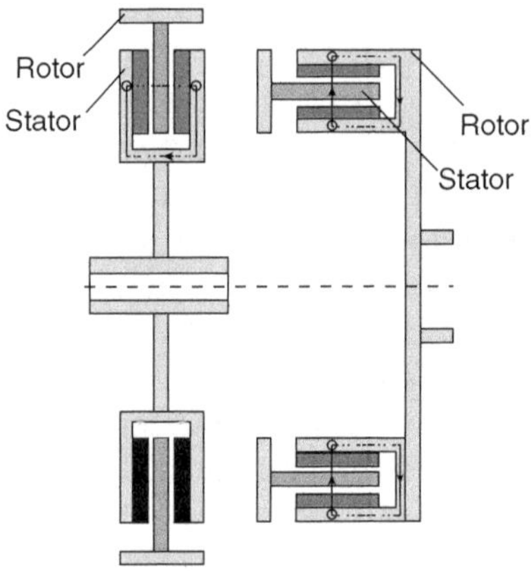

1.18 The air cored PMDD generator topology suggested by Mueller *et al.* AF orientation (left) and RF orientation (right).[8]

stiffness.[84] Relieved from the electromagnetic attraction forces, the structure has only to support its weight relieved from large bending moments. Multistage is also benefited by the C-core structure as all unbalanced loads are successfully transmitted to the main structure. The finite magnetic gap of this design, unlike that in References 37 and 38, proves to be beneficial in terms of the magnetic circuit that enables the maximum airgap flux density with the minimum permanent magnet material. Also the resulting magnetic loading is minimized without having to resort to large diameters like conventional double-sided AF topologies. In addition, the high degree of modularity is beneficial in terms of manufacture, operation and maintenance as the machine can be kept on the grid when a fault occurs in one of the machine's modules or stages without having to stop the rest of the stages and the generator's continuous operation. A 1 MW prototype of such a C-core

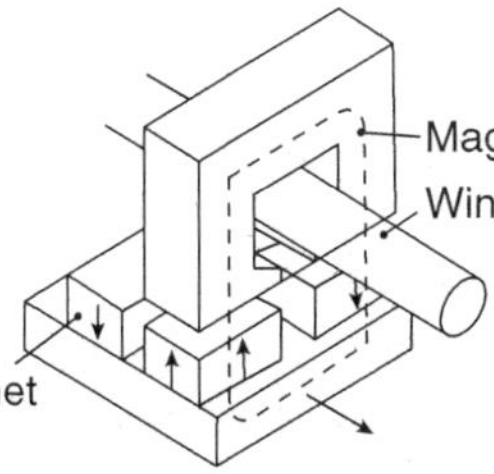

1.19 A single-sided surface-mounted TFPM machine.[95]

air cored AF PMDD generator has been developed and commercialized by NGenTec Ltd, a spin-off company from the University of Edinburgh.[85,86] A multi-MW machine would be possible by adding stages to the machine.

1.3.3 Transverse flux (TF) PMDD machines

In a TF machine the path of the magnetic flux in the core is perpendicular to the direction of the rotor rotation (Fig. 1.19).

TF PMDD topologies allow very small pole pitches (τ_p) to be used leading to higher current loadings and force density compared to other PMDD machine types. Furthermore, they allow an increase in the winding space without decreasing the space available for the main flux. The winding reduces the total amount of copper that is used for these machines, reducing copper losses, allowing lower values of mass/torque ratios to be achieved and minimizing the required active mass.[19] Dubois *et al.* in their extended comparison of a number of suggested direct drive technologies came to the conclusion that the iron cored TF PMDD topology offered the greatest potential in terms of power density and cost/torque ratio.[44]

A fundamental issue for TF PMDD generators is their low power factor (typically between 0.35 and 0.55) due to large armature leakage fields.[87] A lower power factor, even with flux concentration (up to 0.7), creates a high reactive power demand and a small real power output, which would make such generators unattractive for large wind turbine applications. However, power factor correction can be achieved through an active current control of the converter connected to each phase of the TF PMDD generator. Schuttler *et al.* used a DSP controller board for each phase converter to optimize the power factor of a C-core TF PMDD generator with flux concentration using a normalized open-circuit voltage as a current reference signal, giving a power factor of 1.[88] Other optimization methods for the power factor include magnetostatic and transient three-dimensional finite element analysis (FEA) for obtaining the best magnetic circuit to minimize the leakage paths of the machine.[88,89]

Despite their many advantages, TF PMDD machines have complicated structures with complex core designs that are difficult to manufacture and assemble compared to RF and AF PMDD ones. Slotless or air cored TF PMDD machines that would simplify the manufacturing procedure are not possible since a sophisticated core design is required to create a flux path perpendicular to the rotor's movement. A very small airgap is also required for such machines.[90] The small airgap combined with the specialized core designs produce a fluctuation in the normal force applied on the structure and result in noise and vibration. A modified magnetic path geometry can reduce these torque ripples and force fluctuations.[91] Various TF PMDD topologies have been proposed to decrease complexity and ease manufacturing and assembly.

A straightforward way to distinguish the large number of TF PMDD designs suggested from the late 1990s until today is by the presence or absence of flux-concentrating PM poles mounted on the rotor disc. Outer rotor topologies have been described by References 90–92 but they create a heavier and more expensive machine according to Reference 6. Thus, inner-rotor topologies seem to be the best choice for TF PMDD machines, with or without flux concentration. Other suggested topologies of TF PMDD machines without flux concentration include single- or double-sided winding[95,96] and different core designs such as U-core,[95,97] U-core with stator bridges[98–100] and claw pole core (Fig. 1.20).[101] However, all surface-mounted TF PMDD machines that lack flux concentration are plagued by a very low power factor that makes them unsuitable for large wind power applications.[6,87,90,102]

TF PMDD machines with flux concentration are superior to their surface-mounted counterparts in terms of force density and power factor. Suggested stator core topologies are similar to the surface-mounted ones, but with more complicated core designs. A single or double winding has been proposed,[103–110] with the former topology advanced as more efficient and lightweight since the second winding increases outer diameter and structural complexity and leads to larger airgaps that reduce torque density.[6,90] The numerous suggested structures can be distinguished by the shape of their stator core, leading to U-core,[103–105] C-core,[103,106,108] E-core[103,105] and claw-pole designs.[106,109,110] Figure 1.21 illustrates an example of each core design that can be found in the literature.

The claw-pole TF PMDD machine with toothed rotor and flux concentrators proposed by Dubois in Reference 108 simplifies manufacturing while providing the same characteristics as conventional RF PMDD machines regarding mass/torque and cost/torque ratios (Fig. 1.22). However, these ratings increase for diameters larger than 1 m and for torque ratings higher than 10 kNm, therefore such a topology would not be suitable for large offshore wind development.

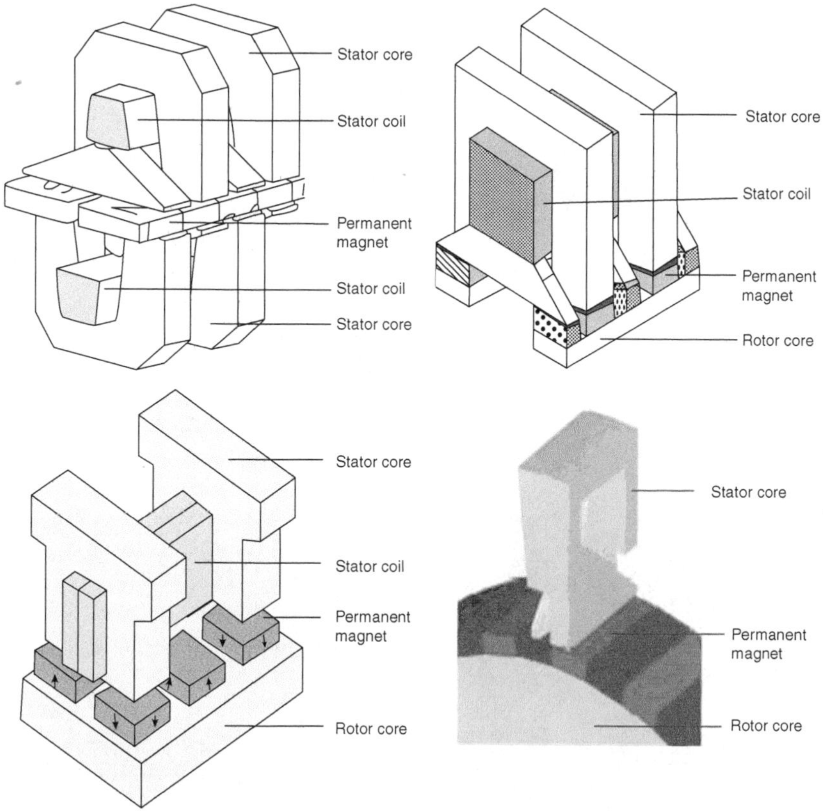

1.20 Suggested TF PMDD topologies found in literature (without flux concentrators).[95,97 98,101]

Bang compared the different surface mounted and flux concentrating TF PMDD designs for their active mass/torque ratio and distinguished the C-core flux concentrating arrangement as the one with the highest torque density for the lightest mass.[19]

1.4 Conclusion

PMDD machines have several advantages, such as increased reliability and higher energy yield, which make them superior to both EEDD and SRG machines. Additionally, the performance of power electronics has been improving and a further reduction in the cost of PM materials is expected in the near future. Therefore, PMDD machines can be more attractive for offshore wind turbine applications.

Slotted RF PMDD machines with an inner rotor and surface-mounted PMs are the lightest and most reliable RF machines.

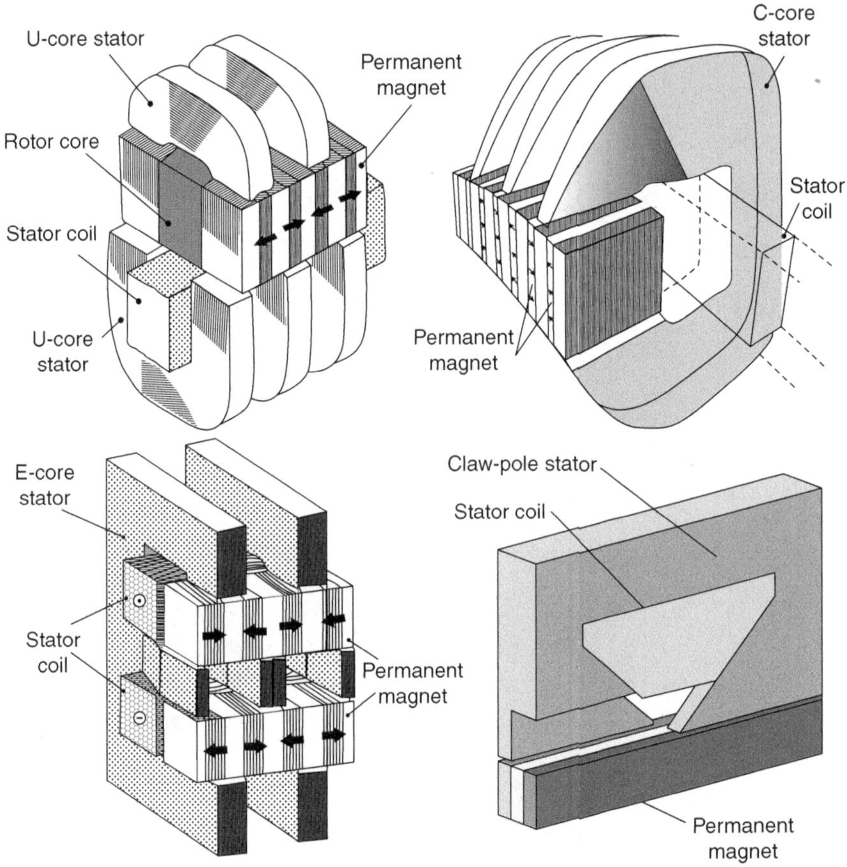

1.21 Suggested TF PMDD topologies found in literature (with flux concentration).[103,106,107]

The C-cored AF PMDD machine suggested in References 8, 82 and 83 provides a structural topology with all the positive attributes of an air cored AF machine – such as structural simplicity, zero cogging torque and increased efficiency – and can effectively maintain the airgap clearance against electromagnetic and structural bending forces, reducing at the same time the structural and active mass requirements compared to other AF or RF PMDD designs.

The C-core TF PMDD generator with an inner rotor and flux concentration was defined as the topology that combines successfully all the beneficial attributes of a TF machine – high force density, high current loadings and reduced copper losses – with the highest torque/mass ratio compared to RF and AF machines or other TF PMDD designs.[19] The same topology was also chosen as the one with the highest potential for taking advantage of novel power factor and structural optimization techniques.[88,89]

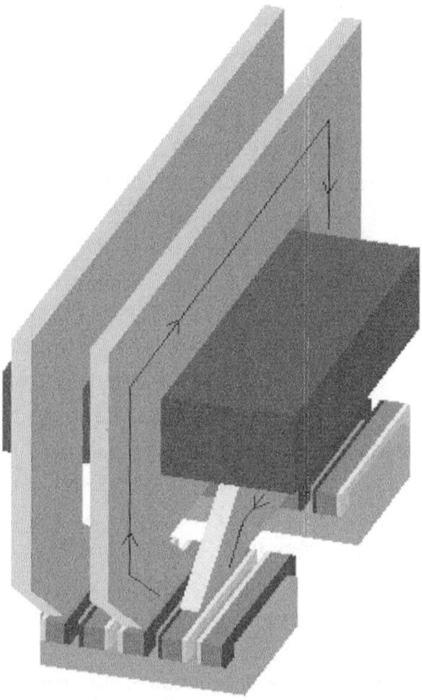

1.22 TF PMDD machine with toothed rotor and flux concentrators.[110]

1.5 References

1. J. R. Harrison, E. Hau and S. Herman, *Large Wind Turbines: Design and Economics*, Wiley, (2000).
2. J. Ribrant and L. M. Bertling, 'Survey of failures in wind power systems with focus on Swedish wind power plants during 1997–2005', *IEEE Transactions on Energy Conversion EC*, Volume **22**, Issue 1, pp. 167–73, (2007).
3. S. Faulstich, B. Hahn and P. J. Tavner, 'Wind turbine downtime and its importance for offshore deployment', *Wind Energy Journal*, Volume **14**, Issue 3, pp. 327–37, April (2011).
4. H. Polinder, 'Overview of and trends in wind turbine generator systems', *Proceedings of IEEE Power and Energy Society General Meeting*, pp. 1–8, July (2011).
5. A.D. Hansen, F. Iov, F. Blaabjerg and L. H. Hansen, 'Review of contemporary wind turbine concepts and their market penetration', *Wind Engineering Journal*, Volume **28**, Issue 3, pp. 247–63, (2004).
6. D. Bang, H. Polinder, G. Shrestha and J. A. Ferreira, 'Comparative design of radial and transverse flux PM generators for direct-drive wind turbines', Electrical Power Processing/DUWIND, Delft University of Technology, International Conference on Electrical Machines, pp. 1–6, Vilamoura, September (2008).
7. A.D. Hansen and L.H. Hansen, 'Wind turbine concept market penetration over 10 years (1995–2004)', *Wind Energy Journal*, Volume **10**, pp. 81–97, (2007).
8. M.A. Mueller, A.S. McDonald and D.E. Macpherson, 'Structural analysis of low-speed axial-flux permanent-magnet machines', *IEE Proceedings of Electric Power Applications*, Volume **152**, No. 6, pp. 1417–26, November (2005).

9. A.S. McDonald, M.A Mueller and H. Polinder, 'Structural mass in direct drive permanent magnet electrical generator', *IET Renewable Power Generation Special Issue – Selected Papers from EWEC 2007*, Volume **2**, No. 1, pp. 3–15, March (2008).
10. G. Bywaters, V. John, J. Lynch, P. Mattila, G. Norton, J. Stowell, M. Salata, O. Labath, A. Chertok and D. Hablanian, 'Northern Power Systems WindPACT drive train alternative design study report', Report NREL/SR-500–35524, (2004).
11. J. Soens, 'Impact of wind energy in a future power grid', Ph.D. thesis, Katholieke Universiteit Leuven, Leuven, Belgium, (2005).
12. H. Polinder, F.F.A. vander Pijl, G.J. de Vilder and P. Tavner, 'Comparison of direct-drive and geared generator concepts for wind turbines', *IEEE Transactions on Energy Conversion*, Volume **21**, pp. 725–33, September (2006).
13. A. Beekmann, J. Marques, E. Quitmann and S. Wachtel, 'Wind energy converters with FACTS capabilities for optimized integration of wind power into transmission and distribution systems', Proceedings of CIGRE/IEEE Integration of Wide-Scale Renewable Resources Into the Power Delivery System Joint Symposium, July (2009).
14. Enercon, www.enercon.de/en-en/, last accessed 18 January 2013.
15. 7-MW-WEC-by-11, www.7mw-wec-by-11.eu, last accessed June 2012.
16. http://www.mtoi.es/en/productos-y-servicios/aerogeneradores-twt.aspx, last accessed July 2012.
17. A. Grauers, 'Design of direct-driven permanent-magnet generators for wind turbines', Ph.D. thesis, Chalmers University of Technology, Göteborg, Sweden, (1996).
18. S. Jöckel, 'Gearless wind energy converters with permanent magnet generators – an option for the future?', *Proceedings of European Union Wind Energy Conference*, pp. 414–17, (1996).
19. D. Bang, 'Design of transverse flux permanent magnet machines for large direct-drive wind turbines', Ph.D. thesis, Delft University of Technology, Delft, The Netherlands, (2010).
20. E. Spooner and B.J. Chalmers, 'TORUS: A slotless, toroidal-stator, permanent magnet generator', International Conference on Electrical Machines, Cambridge, pp. 1053–8, (1990).
21. M. Liserre and M. Molinas, 'Overview of multi-MW wind turbines and wind parks', *IEEE Transactions on Industrial Electronics*, Volume **58**, Issue 4, April (2011).
22. J.M. Carrasco, L.G. Franquelo, J.T. Bialasiewicz, E. Galvan, R.C.P. Guisado, Ma.A.M. Prats, J.I. Leon and N. Moreno-Alfonso, 'Power-electronic systems for the grid integration of renewable energy sources: A survey', *IEEE Transactions on Industrial Electronics*, Volume **53**, Issue 4, June (2006).
23. N. Vilsbøll, A. Pinegin, D. Goussarov and J. Bugge, 'The experience of designing and testing a 20 kW multi pole permanent magnet generator for wind turbines', *DEWI Magazine*, No. 9, pp. 74–83, August (1996).
24. P.J. Tavner, J. Xiang and F. Spinato, 'Reliability analysis for wind turbines', *Wind Energy Journal*, Volume **10**, pp. 1–18, (2007).
25. www.gereports.com/ges-offshore-technology-expands-with-scanwind-buy/, last accessed July 2012.
26. www.ge-energy.com/products_and_services/products/wind_turbines/index.jsp, last accessed July 2012.

27. www.siemens.com/wind, last accessed July 2012.
28. www.goldwindglobal.com, last accessed July 2012.
29. www.stxwind.com/nl/products/7-products, last accessed July 2012.
30. www.ewtinternational.com/products/product-single/dw-9096–20-mw.html, last accessed July 2012.
31. en.leitwind.com/Products-Services/Product-Overview2, last accessed July 2012.
32. www.vensys.de/energy-en/produkte-und-service/produkte-und-service.php, last accessed July 2012.
33. xemc-darwind.com/index.php/solutions.html, last accessed July 2012.
34. openbookproject.net/electricCircuits/AC/AC_13.html, last accessed June 2012
35. D.A. Torrey, 'Variable reluctance generators in wind energy systems', *Proceedings of IEEE Power Electronics Specialists Conference*, pp. 561–77, (1993).
36. M.A. Mueller, 'Design of low speed reluctance machines for wind energy converters', *Proceedings of IEE Electrical Machines and Drives Conference*, Conference Publications No. 468, pp. 60–4, (1999).
37. D.A. Torrey, 'Switched reluctance generators and their control', *IEEE Transactions on Industrial Electronics*, Volume **49**, No. 1, pp. 3–14, February (2002).
38. S.W.H. De Haan, A.T. Veltman, J. Janousek, Z. Cerovsky, J. Mericka, J. Pavelka, J. Perina and F. Petrasek, 'A high efficiency electrical conversion system with variable reluctance generator for variable speed wind turbine', *ECN* report no. ECN-C – 95–033, (1995).
39. J.R. Bumby and R. Martin, 'Axial-flux permanent-magnet air-cored generator for small-scale wind turbines', *Electric Power Applications, IEE Proceedings*, Volume **152**, Issue 5, September (2005).
40. E. Spooner, P. Gordon, J.R. Bumby and C.D. French, 'Lightweight, ironless-stator, PM generators for direct-drive wind turbines', *IEE Proceedings of Electrical Power Applications*, Volume **152**, No. 1, pp. 17–26, January (2005).
41. E. Spooner and A.C. Williamson, 'Direct coupled, permanent magnet generators for wind turbine applications', *IEE Proceedings of Electrical Power Applications*, Volume **143**, No. 1, pp. 1–8, January (1996).
42. E. Spooner and A.C. Williamson, 'Modular, permanent magnet wind turbine generators', *Proceedings of IEEE Industry Applications Conference*, Volume **1**, pp. 497–502, (1996).
43. E. Spooner, A.C. Williamson and G. Catto, 'Modular design of permanent-magnet generators for wind turbines', *IEE Proceedings of Electrical Power Applications*, Volume **143**, No. 5, pp. 388–95, September (1996).
44. M.R. Dubois, H. Polinder and J.A. Ferreira, 'Comparison of generator topologies for direct-drive for wind turbines', Nordic Countries Power Industrial Electronics Conference Proceedings, Aalborg, Denmark, pp. 22–6, June (2000).
45. E. Spooner and A.C. Williamson, 'Permanent-magnet generators for wind turbine applications', *Proceedings of IEEE Electrical Power Applications*, Volume **143**, No. 1, pp. 1–8, (1996).
46. J. Chen, C.V. Nayar and L. Xu, 'Design and finite-element analysis of an outer-rotor permanent-magnet generator for directly coupled wind turbines', *IEEE Transactions on Magnetics*, Volume **36**, No. 5, pp. 3802–9, September, (2000).

47. F. Libert and J. Soulard, 'Design study of different direct-driven permanent-magnet motors for a low speed application', in Proceedings of the Nordic Workshop on Power and Industrial Electronics (NORpie), Trondheim, Norway, June (2004).
48. F. Libert and J. Soulard, 'Design study of low-speed direct-driven permanent magnet motors with concentrated windings', Proceedings of 6th International Symposium on Advanced Electromechanical Motion Systems (Electromotion), September (2005).
49. W. Wu, V.S. Ramsden, T. Crawford and G. Hill, 'A low-speed, high-torque, direct drive permanent magnet generator for wind turbine', *Proceedings of IEEE Industrial Applications Conference*, Volume **1**, pp. 147–54, (2000).
50. G. Korouji, 'Design, construction and test study of a wind energy generator with dual permanent magnet excitation', Ph.D. thesis, Berlin University of Technology, Berlin, Germany, (2004).
51. M.S. Widyan, 'Design, optimization, construction and test of rare-earth permanent-magnet electrical machines with new topology for wind energy applications', Ph.D. thesis, Berlin University of Technology, Berlin, Germany, (2006).
52. H. Polinder, M.J. Hoeijmakers and M. Scuotto, 'Eddy-current losses in the solid back-iron of permanent-magnet machines with different concentrated fractional pitch windings', *Proceedings of IEEE International Electrical Machines and Drives Conference (IEMDC07), Antalya*, pp. 652–7, 3–5 May (2007).
53. E. Spooner, P. Gordon, J.R. Bumby and C.D. French, 'Lightweight, ironless-stator, PM generators for direct-drive wind turbines', *IEE Proceedings of Electrical Power Applications*, Volume **152**, No. 1, pp. 17–26, January (2005).
54. C.J.A. Versteegh, 'Design of Zephyros Z72 wind turbine with emphasis on the direct drive PM generator', *Proceedings of Nordic Workshop Power Industrial Electronics (NORPIE)*, Paper No. 68, Trondheim, Norway, 14–16 June 2004.
55. S. Engstrom and S. Lindgren : 'Design of NewGen direct drive generator for demonstration in a 3.5 MW Wind Turbine', EWEC, Milan, Italy, (2007).
56. G. Shrestha, H. Polinder, D.J. Bang and J.A. Ferreira, 'Structural flexibility: A solution for weight reduction of large direct-drive wind-turbine generators', *IEEE Energy Conversion Journal*, Volume **25**, Issue 3, pp. 732–40, September (2010).
57. G. Shrestha, H. Polinder, D.J. Bang and J.A. Ferreira, 'Direct drive wind turbine generator with magnetic bearing', EOW conference, Berlin (2007).
58. J.N. Stander, G. Venter and M.J. Kamper, 'Review of direct-drive radial flux wind turbine generator mechanical design', *Wind Energy Journal*, Volume **15**, Issue 3, July (2011).
59. P. Campbell, 'Principles of a permanent-magnet axial-field DC machine', *Proceedings of IEE*, Volume **121**, No. 12, pp. 1489–94, December (1974).
60. P. Campbell, 'The magnetic circuit of an axial-field DC electrical machine', *IEEE Transactions on Magnetics*, Volume **11**, No. 5, pp. 1541–3, September (1975).
61. G. Henneberger, H. Harer, S. Schustek and L. Verstege, 'A new range of DC and AC pancake motors', *Proceedings of International Conference on Electrical Machines (ICEM)*, pp. 916–19, (1986).
62. H. Weh, 'High power synchronous machines with permanent magnet excitation', *Proceedings of International Conference on Electrical Machines (ICEM)*, pp. 295–303, (1980).
63. C. C. Chan, 'Axial-field electrical machines-design and applications', *IEEE Transactions on Energy Conversion*, Volume EC-**2**, No. 2, pp. 294–300, (1987).

64. G. B. Kliman, 'Permanent magnet AC disc motor electric vehicle drive', SAE Technical Paper Series, International Congress and Exposition, Detroit, Michigan, (1983).
65. Z. Zhang, F. Profumo and A. Tenconi, 'Axial-flux versus radialflux PM motors', *SPEEDAM*, Italy, pp. A4–19–25, (1996).
66. J. Rizk and M. Nagrial, 'Performance of axial type coupling', Proceedings of International Conference on Electrical Machines (ICEM), (1998).
67. S. Huang, M. Aydin and T.A. Lipo, 'Torque quality assessment and sizing optimization for surface mounted PM machines', *IEEE Industry Applications Society Annual Meeting*, pp. 1603–10, Chicago, September (2001).
68. A. Cavagnino, M. Lazzari, F. Profumo and A. Tenconi, 'A comparison between the axial flux and the radial flux structures for PM synchronous motors', *IEEE Transactions on Industry Applications*, Volume **38**, No. 6, pp. 1517–24, November/December (2002).
69. K. Sitapati and R. Krishnan, 'Performance comparison of radial and axial field permanent magnet brushless machines', *IEEE Transactions on Industry Applications*, Volume **37**, No. 5, pp. 1219–26, September/October (2001).
70. Y. Chen, P. Pillay and A. Khan, 'PM wind generator topologies', *IEEE Transactions on Industrial Applications*, Volume **41**, Issue 6, pp. 1619–26, (2005).
71. M. Aydin, S. Huang and T.A. Lipo, 'Axial flux permanent magnet disc machines: A review', Symposium on Power Electronics, Research Report, (2004).
72. E. Spooner and B.J. Chalmers, 'TORUS, a slotless, toroidal stator, permanent magnet generator', *Proceedings of IEE, Part-B*, Volume **139**, No. 6, pp. 497–506, November (1992).
73. E. Spooner and B.J. Chalmers, 'Toroidally-wound, slotless, axial-flux, permanent-magnet, brushless-DC motors', *Proceedings of International Conference on Electrical Machines (ICEM)*, pp. 81–6, (1988).
74. B.J. Chalmers, A.M. Green, A.B.J. Reece and A.H. Al-Badi, 'Modeling and simulation of the TORUS generator', *IEE Proceedings on Electric Power Applications*, Volume **144**, No. 6, pp. 446–52, November (1997).
75. B.J. Chalmers, W. Wu and E. Spooner, 'An axial flux permanent-magnet generator for a gearless wind energy system', *IEEE Transactions on Energy Conversion*, Volume **14**, No. 2, pp. 251–7, (1999).
76. R.J. Wang, M.J. Kamper, K. VanderWesthuizen and J.F. Gieras, 'Optimal design of a coreless stator axial flux permanent-magnet generator', *IEEE Transactions on Magnetics*, Volume **141**, Issue 1, pp. 55–64, (2005).
77. N.F. Lombard and M.J. Kamper, 'Analysis and performance of an ironless stator axial flux PM machine', *IEEE Transactions on Energy Conversion*, Volume **14**, Issue 4, pp. 1051–6, (1999).
78. F. Caricchi, F. Crescimbini, F. Mezzetti and E. Santini, 'Multi sage axial flux PM machine for wheel direct drive', *IEEE Industry Applications Society Annual Meeting*, pp. 679–84, (1995).
79. F. Caricchi, F. Crescimbini, F. Mezzetti and ESantini, 'Multistage axial-flux PM machine for wheel direct drive', *IEEE Transactions on Industry Applications*, Volume **34**, No. 4, pp. 882–8, July/August (1996).
80. E. Spooner and A.C. Williamson, 'Direct coupled, permanent magnet generators for wind turbine applications', *Proceedings of IEE Electric Power Application*, Volume **143**, No. 1, pp. 1–8, January (1996).

81. R.J. Hill-Cottingham, P.C. Coles, J.F. Eastham, F. Profumo, A. Tenconi and G. Gianolio, 'Multi-disc axial flux stratospheric aircraft propeller drive', IEEE Industry Applications Society Annual Meeting, (2001).
82. J. Braid, A. van Zyl and C. Landy, 'Design, analysis and development of a multistage axial-flux permanent magnet synchronous machine', *Proceedings of 6th IEEE Africon Conference*, Africa, Volume **2**, pp. 675–80, October (2002).
83. J. Braid, A. van Zyl and C. Landy, 'Unbalanced load sharing in a prototype multistage axial-flux permanent magnet synchronous machine', *IEEE International Conference on Electric Machines and Drives 2003 (IEMDC'03)*, Volume **3**, pp. 1935–40, 1–4 June (2003).
84. M.A. Mueller and A.S. McDonald, 'A lightweight lowspeed permanent magnet electrical generator for direct drive wind turbines', *Journal of Wind Energy*, Volume **12**, Issue 8, pp. 768–80, November (2009).
85. A.S. McDonald, N. Al-Khayat, D. Belshaw, M. Ravilious, A. Kumaraperumal, M. Benatmane, M. Galbraith, D. Staton, K. Benoit and M. Mueller, '1 MW multi-stage air-cored permanent magnet generator for wind turbines', *IET Power Electronics Machines and Drives Conference (PEMD)*, Bristol, pp. 27–29 March (2012).
86. NGenTec, www.ngentec.com/, last accessed July 2012.
87. M.R. Harris, G.H. Pajooman and S.M. Abu Sharkh, 'The problem of power factor in VRPM (transverse-flux) machines', *Eighth International Conference on Electrical Machines and Drives* (Conf. Publ. No. 444), pp. 386–90, (1997).
88. J. Schüttler, H. Groke, M. Siatkowski, J. Adler and B. Orlik, 'Power optimized symmetrizing current control with a 8.7 kNm transverse flux generator', *Proceedings of 12th International Conference on Optimization of Electrical and Electronic Equipment (OPTIM)*, pp. 352–7, (2010).
89. S. Hosseini, J.S. Moghani, N.F. Ershad and B.B. Jensen, 'Design, prototyping, and analysis of a novel modular permanent-magnet transverse flux disk generator', *IEEE Transactions on Magnetics*, Volume **47**, Issue 4, pp. 772–80, April (2011).
90. M.R. Dubois and H. Polinder, 'Study of TFPM machines with the toothed rotor applied to direct-drive generators for wind turbines', Proceedings of Nordic Workshop on Power and Industrial Electronics, Trondheim, Norway, June (2004).
91. M. Vinogradski, U. Werner and B. Orlik, 'Genetic algorithms used for geometrical structure design of transverse flux permanent magnet motors to optimize the torque wave form', PCIM Nürnberg, Germany, (2004).
92. K.Y. Lu, E. Ritchie, P.O. Rasmussen and P. Sandholdt, 'Modelling a single phase surface-mounted permanent magnet transverse flux machine based on Fourier Series Method', *Proceedings of IEEE Conference on Electric Machines and Drives*, Volume **1**, pp. 340–5, (2003).
93. Y.G. Guo, J.G. Zhu, P.A. Watterson and W. Wu, 'Development of a PM transverse flux motor with soft magnetic composite core', *IEEE Transactions on Energy Conversion*, Volume **21**, No. 2, pp. 426–34, June (2006).
94. W.M. Arshad, T. Bäckström and C. Sadarangani, 'Investigating a transverse flux machine with intermediate poles', *Proceedings of IEE Conference on Power Electronics, Machines and Drives (PEMD)*, pp. 325–8, (2002).
95. H. Weh, H. Hoffmann, J. Landrath, H. Mosebach and J. Poschadel, 'Directly driven permanent magnet excited synchronous generator for variable speed operation', *European Community Wind Energy Conference*, pp. 566–72, (1988).

96. P. Anpalaham, 'Design of transverse flux machines using analytical calculations and finite element analysis', Tech. Licentiate Thesis, Royal Institute of Technology, Stockholm, (2001).
97. D. Svechkarenko, J. Soulard and C. Sadarangani, 'A novel transverse flux generator in direct-driven wind turbines', Proceedings of Nordic Workshop on Power and Industrial Electronics, June (2006).
98. G. Henneberger and M. Bork, 'Development of a new transverse flux motor', in *Proceedings of IEE Colloquium on New Topologies for PM Machines*, Volume **1**, pp. 1–6, (1997).
99. G. Kastinger and A. Schumacher, 'Reducing torque ripple of transverse flux machines by structural designs', *Proceedings of IEE Conference on Power Electronics, Machines and Drives (PEMD)*, pp. 320–4, (2002).
100. J.F. Gieras, 'Performance characteristics of a permanent magnet transverse flux generator', *Proceedings of IEEE Conference on Electric Machines and Drives*, pp. 1293–9, (2005).
101. P. Dickinson, A.G. Jack and B.C. Mecrow, 'Improved permanent magnet machines with claw pole armatures', *Proceedings of International Conference on Electric Machines*, paper **245**, (2002).
102. M.R. Harris, G.H. Pajooman, S.M.A. Sharkh and B.C. Mecrow, 'Comparison of Flux-Concentrated and Surface-Magnet Configurations of the VRPM (Transverse-Flux) machine', *Proceedings of the International Conference on Electrical Machines*, p. 1119, (1998).
103. H. Weh, 'Transverse-flux machines in drive and generator application', *Proceedings of the IEEE Symposium on Electric Power Engineering (Stockholm Power Tech), Strockholm, Sweden*, Volume Invited speaker' session, pp. 75–80, (1995).
104. E. Schmidt, '3-D finite element analysis of the cogging torque of a transverse flux machine', *IEEE Transactions on Magnetics*, Volume **41**, No. 5, pp. 1836–9, May (2005).
105. B.E. Hasubek and E.P. Nowicki, 'Design limitations of reduced magnet material passive rotor transverse flux motors investigated using 3D finite element analysis', *Proceedings of IEEE Canadian Conference on Electrical and Computer Engineering*, Volume **1**, pp. 365–9, (2000).
106. C.P. Maddison, 'Transverse flux machines for high torque applications', Ph.D. thesis, University of Newcastle upon Tyne, UK, (1999).
107. A.J. Mitcham, 'Transverse flux motors for electric propulsion of ships', *Proceedings of IEE Colloquium on New Topologies for PM Machines*, Volume **3**, pp. 1–6, (1997).
108. Y. Rang, Chenglin Gu and H. Li, 'Analytical design and modeling of a transverse flux permanent magnet machine', *Proceedings of Power System Technology Conference*, Volume **4**, pp. 2164–7, (2002).
109. A. Masmoudi, A. Njeh, A. Mansouri, H. Trabelsi and A. Elantably, 'Optimizing the overlap between the stator teeth of a claw pole transverse-flux permanent-magnet machine', *IEEE Transactions on Magnetics*, Volume **40**, No. 3, pp. 1573–8, May (2004).
110. M. Dubois, 'Optimized permanent magnet generator topologies for direct drive wind turbines', Ph.D. thesis, Delft University of Technology, Delft, The Netherlands, (2004).

2

Principles of electrical design of permanent magnet generators for direct drive renewable energy systems

H. POLINDER, Delft University of Technology, The Netherlands

DOI: 10.1533/9780857097491.1.30

Abstract: The objective of this chapter is to discuss the electromagnetic design of direct drive generators for renewable energy generation. It starts with a look at the requirements of these generator systems. Next, it discusses the main material choices and topology choices with their advantages and disadvantages. Then, as an example, the electromagnetic design of a 3 MW 15 rpm permanent magnet (PM) generator for a direct drive wind turbine is discussed. Next, it gives scaling laws that make it possible to estimate the size of direct drive generators. Finally, a number of issues that need further investigation are discussed.

Key words: renewable energy, wind energy, permanent magnet generators, electrically excited generators, design choices, design optimization, scaling.

2.1 Introduction

Direct drive generator systems for renewable energy conversion have been introduced in the first chapter. This second chapter focuses on the electromagnetic design of direct drive generators, There are a number of important design choices that have to be made, and the engineer then needs to decide which methods should be used to design an optimal direct drive generator. In this chapter, potential solutions to these decisions will be presented.

2.2 Design requirements and evaluation criteria

A direct drive generator system has to convert an input force or torque and speed into electrical power without using a gearbox. The system has to meet many requirements, including grid connection and services, manufacture, transport and installation. Polinder *et al.* (2005a) and Polinder (2011) review different generator systems for wind turbines.

The grid imposes requirements on grid voltage, grid frequency, grid fault ride through and power factor, all of which will be discussed in later chapters. In this chapter, it is assumed the grid connection is made via a power electronic converter that has full control of the currents in the generator, as is the case when a voltage source inverter is used. The use of such a converter also makes variable speed control possible.

There are requirements regarding the manufacture, transportation and installation of the generator: what diameter and mass can the generator manufacturer deal with, and what are the limitations of transportation and installation? Direct drive generators are so large and heavy that there are often many practical limitations, and it is wise to consider them at an early stage of the design. Since these considerations vary widely in different situations, however, they will not be discussed further in this chapter.

There are also requirements related to the location where the generator is used, because the environment may affect the life of the generator. Salt and humidity have a detrimental effect on insulation materials and permanent magnets (PMs) if not well protected. Although this is not the main focus of this chapter, it is important to note this issue.

In this chapter, we focus on the requirements from the input side of the generator:

1. The most important requirements are the torque or force and the speed of the generator.
2. In many cases, the cogging torque has to be sufficiently small to enable startup at low power levels and obtain acceptable levels of audible noise.
3. High efficiency may also be listed as a requirement, but in practice the efficiency is mostly the result of an optimization.

In addition to listing the design requirements, exploring the criteria used to evaluate the design may help with design choices. The most obvious criterion is cost: the cheapest generator that does the job is the best one. However, at least two further questions need to be considered. Firstly, what costs are included? Material cost is the most easy to evaluate, but there is also manufacturing cost, transportation cost, operation and maintenance cost and decommissioning cost. Secondly, what is the precise job that the generator is required to do? Most renewable energy systems operate at part load for a significant part of their life, so a high part-load efficiency may significantly increase the energy yield. Taking everything into account gives the best optimization criterion, but it may be difficult to determine all the important parameters, particularly those related to operation and maintenance.

2.3 Scaling laws for dimensioning machines

Before looking at some of the details of electromagnetic design, it is useful to outline the basic design principles for sizing machines. A good starting point for the sizing equations of machines is the force density F_d (the force acting per unit surface area of the air gap) illustrated in Fig. 2.1 (Miller, 1989; Grauers & Kasinathan, 2004; Grauers *et al.*, 2005):

$$F_d \approx \frac{1}{2}\hat{B}_g\hat{A}_s\cos(\gamma) \qquad [2.1]$$

where

$\hat{B}_g$ is the amplitude of the fundamental of the air-gap flux density due to the magnets,

$\hat{A}_s$ is the amplitude of the fundamental of the surface current density of the stator (in A/m), and

γ is angle between the maxima of the flux density and the linear current density.

This force density is rather constant for very different machines, because the air-gap flux density is limited by saturation of the stator teeth, and the linear current density is limited by the maximum allowed heat dissipation.

Practical values for the force density of air-cooled direct drive machines are in the range of $F_d = 30 - 60\text{kN/m}^2$, depending on the cooling methods (Ruuskanen *et al.*, 2011). For liquid cooled machines higher values may be possible.

Using this force density, the power P produced by a machine can be written as

$$P = \omega_m T = \omega_m r_s F = 2\omega_m \pi r_s^2 l_s F_d = 2\omega_m V_r F_d \qquad [2.2]$$

where

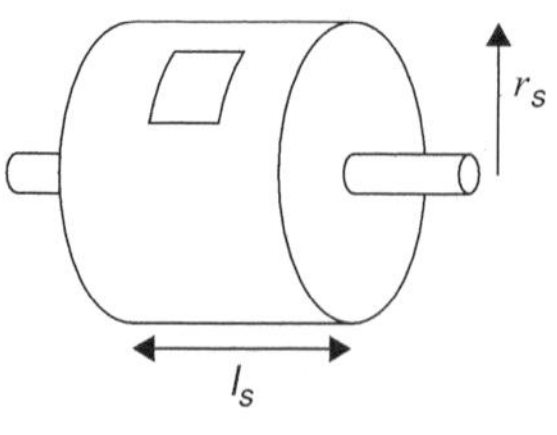

2.1 The rotor surface with a surface area that produces a force.

ω_m is the mechanical angular speed of the rotor;
T is the torque,
r_s is the air-gap radius of the machine
l_s is the stack length of the machine, and
V_r is the volume of the rotor.

This equation shows clearly why direct drive machines are large: if speed is low, the volume of the rotor has to be large to enable a high power level.

2.4 Design choices

In a PM machine the main materials are laminated electrical steel, copper and PMs. The grade of material used and the relative amounts of each material impact the design of the machine.

2.4.1 Material choices

Laminated electrical steel

The laminations are used to provide a low reluctance path for the magnetic flux particularly in regions where the flux density changes. The types of electrical steel used are characterized by the saturation flux density and the iron losses. The saturation flux density should be as high as possible because that gives the highest possible force density.

Generally, laminations with a higher saturation level have a lower percentage of silicon and higher iron losses, so there is a trade-off between saturation flux density and losses. The frequencies in direct drive machines are generally so low that at rated load the iron losses are small compared to the copper losses. However, at partial load the iron losses can be significant.

Copper for windings

Copper windings make it possible to extract electrical power from the generator. There are a number of options available in winding design, including single-layered, double-layered, concentrated, round wire coils, strip wound coils and preformed coils.

The winding arrangement is driven by the application. It is beyond the scope of this book to cover winding design. The reader is directed to references Richter (1967) and Fitzgerald *et al.* (2003) for more detailed design information. Round copper is cheaper than diamond coils, but the fill factor of the slot is also lower (typically 40% compared to 70%), which results in higher copper losses.

Round copper is mostly used in single-layer windings. If preformed coils are used, they could form a single-layer winding or a two-layer winding. In

Table 2.1 Indications for characteristics of permanent magnets

	B_r (T)	H_c (kA/m)	dB_r/dT (%/K)	dH_c/dT (%/K)	ρ (μΩm)	Price (€/kg)
NdFeB	1.44	1100	−0.2	−0.3	1.3	100
SmCo	1	700	−0.05	−0.3	0.5	200
Ferrite	0.4	250	−0.2	+0.3	Inf	5

a two-layer winding, preformed coils are normally called diamond coils. In this case, it is possible to short pitch the winding in order to reduce harmonics in the machine.

It is very important to choose the proper insulation and impregnation techniques. There is a lot of thermal cycling because the generator heats up at high loads and cools down at low loads. In combination with humidity and salt, this may damage the winding and lead to short circuits.

The maximum allowable current density in direct drive generators depends on cooling: insulation materials deteriorate at temperatures higher than rated. For direct drive machines with natural air cooling, root mean square (RMS) values of the current density in the order of 3–4 A/mm^2 are typical. For direct drive machines with liquid cooling systems, higher values are possible.

Magnets

The most commonly used type of PM is the neodymium iron boron (NdFeB) magnet, a so-called rare earth magnet. It became the most commonly used type of magnet because of the combination of a high magnetic field and a low price. In the period 1995–2005, the cost fell by a factor of 10 or so. However, the increasing demand for PMs combined with measures to protect the market in China led to peak magnet prices in 2011, when the price increased by a factor of 8 or so. Since then, the magnet price has fallen again, but there is still a lot of uncertainty around future price trends.

PMs are defined by their BH characteristic. Because the BH characteristic of the most commonly used magnets is a straight line in the second quadrant, they can also be characterized by the remanent flux density B_r and the coercive force H_c (see Table 2.1).

When PMs are used, there are a number of issues that have to be considered:

1. The resistivity of PMs is not infinite, so there may be eddy-current losses in the generator, especially when concentrated fractional pitch windings are used. These losses may be acceptable if the cooling is sufficient to remove the heat.

2. The magnetization decreases reversibly when the temperature increases, so the magnetization at the maximum operating temperature should be used in the design of the machine.
3. Permanent demagnetization has to be avoided, because if the PMs lose their magnetization, the generator becomes useless.

There are three possible causes of demagnetization:

1. The remanent flux density and the coercive force depend on temperature, so magnets may be demagnetized by a high temperature. These high temperatures must be avoided.
2. High currents that make a magnetic field in the direction opposite to the direction of magnetization. These currents may occur during a short circuit of the generator. Therefore, generators are normally designed in such a way that the magnets do not demagnetize in case of an unexpected short circuit, even when the generator has reached its maximum operating temperature.
3. NdFeB magnets are easily corroded. Therefore, especially in aggressive (salty and humid) environments, they have to be protected extremely well. Sometimes, they are completely enclosed in a metal shield.

2.4.2 Machine topology choices

Radial-flux or axial-flux generators

Most direct drive wind turbines have radial-flux generators. Jeumont is the only company to have built direct drive wind turbines with axial-flux machines. Several people investigated axial-flux generators (Lampola, 2001). Although axial-flux machines may be smaller than radial-flux machines, they have the following disadvantages (Dubois, 2004):

- The force density is not optimal at all radii; because the slots have a constant width, the teeth are narrow and probably heavily saturated at the inner radius, while the teeth are wide and probably not yet saturated at the outer radius.
- Production is more difficult because the laminations are different at different radii.
- Attractive forces between stator and rotor increase the bearing loads.

Electrical or permanent magnetic (PM) excitation

Figures 2.2 and 2.3 depict a few pole pitches of a machine with electrical excitation and of a machine with PM excitation. The successful direct drive

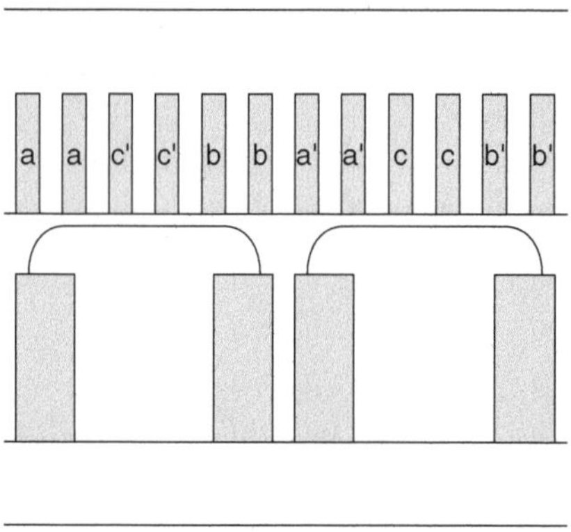

2.2 Cross-section of two pole pitches of an electrically excited generator.

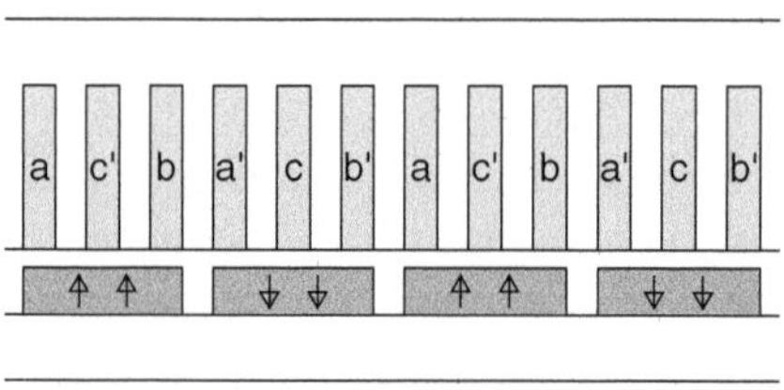

2.3 Cross-section of four pole pitches of a permanent magnet excited generator.

manufacturer Enercon uses direct drive generators with electrical excitation. Most of the other companies working on direct drive generators started more recently (Siemens, General Electric) and use PM excitation. The advantages of electrical excitation are the following:

- There is no risk of demagnetization of the magnets.
- There are no problems with the availability of rather expensive rare earth PMs.
- There is the additional possibility of controlling the field current, which can be reduced to reduce the losses at low wind speeds to enable very low cut-in wind speeds, and to reduce the induced voltage during high rotational speeds so that it is not necessary to overrate the converter.

The advantages of permanent magnet excitation are the following (Grauers, 1996a, 1996b; Hartkopf *et al.*, 1997; Dubois, 2004; Joeckel *et al.* 2006; Polinder *et al.*, 2006):

- There are no excitation losses, which on average leads to a better energy yield.

- The weight of active materials can be significantly reduced, because PMs can be much smaller than poles with electrical excitation.
- According to Polinder *et al.* (2006), PM machines can be cheaper than electrically excited generators.

Stators with slots, air-gap winding or air cored machines

All direct drive wind turbine manufacturers use a stator with a winding in slots. Spooner *et al.* (2005) and McDonald *et al.* (2008) proposed the use of air-gap windings or even air cored machines (without iron in the stator) to avoid the large attractive forces between stator and rotor and therefore to enable light constructions. Manufacturers are probably sticking to the well-known technology of direct drive machines with windings in slots, because they consider these machines to be more robust, more efficient and cheaper.

Open slots, semi-closed slots or slots with magnetic wedges

Windings are mostly placed in slots. Slot wedges are used to prevent the windings from coming out of the slots during operation. In order to have a minimum reluctance for the main flux due to the magnets, the slot opening should be small. This also helps to reduce cogging and noise. However, a minimum size of slot opening is necessary for insertion of the windings. For round copper, this slot opening can be quite small. For diamond coils, the slot opening has to be as wide as the rest of the slot.

Distributed windings or concentrated fractional pitch windings

Most classic AC machines have distributed windings: the number of slots per pole per phase is one (as in Figs 2.2 and 2.3) or more than one. They can be wound as a single-layer or a two-layer winding. In a two-layer winding, preformed coils or diamond coils are used. In this case, it is possible to short pitch the winding in order to reduce harmonics in the machine (Richter, 1967; Fitzgerald *et al.*, 2003).

However, fractional pitch concentrated windings as illustrated in Fig. 2.4 are increasingly used in PM machines (Cros & Viarouge, 2002; Polinder *et al.*, 2003; El-Refaie *et al.*, 2006; Wu *et al.*, 2010). In these machines, the number of stator teeth is generally quite similar to the number of rotor poles. This has the following advantages:

- The end windings can be very short, which makes the machine more compact.
- The stator yokes can be thinner because the pole pitches are mostly smaller.
- The risk of demagnetizing the magnets is smaller. The manufacturing cost may be significantly lower.

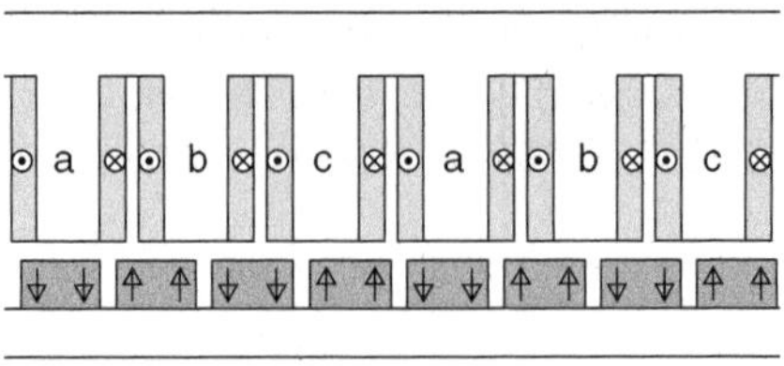

2.4 Cross-section of a machine with a concentrated fractional pitch winding.

However, these machines have significant space harmonics (in most cases including harmonics with a wave length longer than twice the pole pitch). This may lead to problematic eddy-current losses in the magnets and the back-iron (Polinder *et al.*, 2007). Furthermore, these machines have unbalanced magnetic forces (Wu *et al.*, 2010). Therefore, most direct drive generators still have distributed windings.

Skew or not

Stator and rotor may be skewed with respect to each other. The magnets on the rotor may be skewed or the stator slots may be skewed. Skewing is a method used to reduce cogging (the existence of preferred rotor positions) and to reduce audible noise.

Buried magnets with flux concentration or surface-mounted magnets

In many machines, surface-mounted magnets are used: magnets are glued or screwed to the back-iron with the magnetization perpendicular to the surface of the rotor, sometimes with an additional glass fiber bandage. However, it is also possible to use buried magnets: magnets that are 'buried' in the iron structure of the rotor. This has the advantage that it is possible to use flux concentration: the flux density in the air gap becomes larger than the flux density in the magnet.

Voltage level selection

In the generator design, it is possible to adapt the voltage level by changing the number of series- and parallel-connected coils in the machine. Mostly, wind turbine manufacturers use standard industrial voltage levels, so that more or less off-the-shelf power electronic converters can be used for the grid connection.

Power factor

Figure 2.5 depicts the equivalent circuit of a synchronous machine and a phasor diagram. The angle between the current and the electromotive force

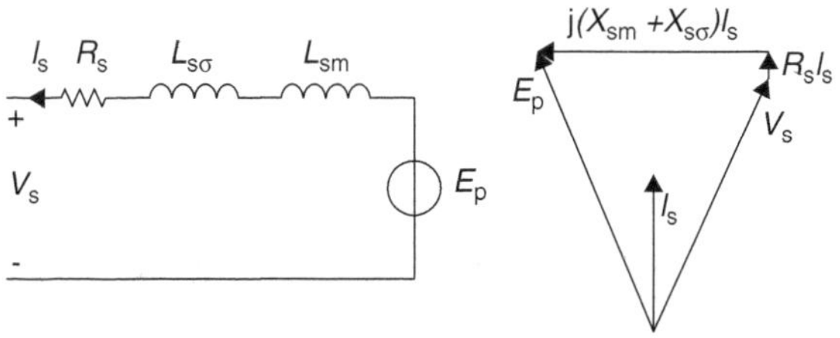

2.5 Equivalent circuit of a PM synchronous machine and phasor diagram where the current phasor I_s is in between the terminal voltage phasor V_s and the electromotive force E_p.

or the angle between the current and the terminal voltage can be chosen. If the current is in phase with the electromotive force, the copper losses are the lowest. However, the flux density level in the machine is increased, which may result in saturation, and the converter has to be overrated because of the low converter power factor. If the current is in phase with terminal voltage, the converter power level can be the smallest. However, this limits the power that can be taken from the generator if the synchronous inductance is large. Therefore, the phase of the current may be set at a point between the phase of the electromotive force and the terminal voltage as indicated in the phasor diagram of Fig. 2.5.

2.5 Design example

As an example, the PM direct drive generator presented in Polinder *et al.* (2006) is discussed. This generator is for a 3 MW wind turbine with a rated speed of 15 rpm.

Based on the discussion in Section 2.3, the following design choices have been made:

- A PM radial-flux generator with surface-mounted magnets is used.
- A three-phase distributed winding with one slot per pole per phase is used.
- A two-layer winding with diamond coils is used.
- Stator slots and magnets are skewed over one slot pitch with respect to each other.
- The generator is loaded with an active rectifier, which keeps the phase of the stator current between the electromotive force and the terminal voltage.

Most of the permanent magnet direct drive generators are of this type (Polinder *et al.*, 2004, 2005b; Versteegh, 2004).

This section describes first how the main parameters of the machine (the parameters of the equivalent circuit) and the main losses are determined, using simple analytical methods. The resulting design is then described.

2.5.1 Parameter calculations

The machine parameters are calculated in conventional ways (Richter, 1967). The following assumptions are used in the calculations:

- Space harmonics of the magnetic flux density distribution in the air gap are negligible; only the fundamental is considered.
- The magnetic flux density crosses the air gap perpendicularly.

The effective air gap of the machine is an important factor for determining the flux densities and the inductances. This effective air gap can be written as

$$g_{\text{eff}} = k_C\left(g + \frac{l_m}{\mu_{\text{rm}}}\right) \quad [2.3]$$

where

k_{C} is the Carter factor (Richter, 1967),
g is the mechanical air gap,
μ_{rm} is the relative recoil permeability of the magnets, and
l_{m} is the magnet length in the direction of the magnetization.

The calculation of the Carter factor within the expression for the effective air gap is modified because of the surface-mounted magnets (Richter, 1967) and given by

$$\begin{aligned} k_{\text{C}} &= \frac{\tau_s}{\tau_s - g_1\gamma} \\ g_1 &= g + \frac{l_m}{\mu_{\text{rm}}} \\ \gamma &= \frac{4}{\pi}\left(\frac{b_{\text{so}}}{2g_1}\arctan\left(\frac{b_{\text{so}}}{2g_1}\right) - \log\sqrt{1+\left(\frac{b_{\text{so}}}{2g_1}\right)^2}\right) \end{aligned} \quad [2.4]$$

where

τ_{s} is the slot pitch, and
b_{so} is the slot opening width.

Using this effective air gap, the fundamental space harmonic of the magnetic flux density in the air gap due to the magnets can be calculated as

$$\hat{B}_g = \frac{l_m}{\mu_{\text{rm}} g_{\text{eff}}} B_{\text{rm}} \frac{4}{\pi} \sin\left(\frac{\pi b_p}{2\tau_p}\right) \qquad [2.5]$$

where

B_{rm} is the remanent flux density of the magnets,
τ_p is the pole pitch,
b_p is the width of the magnet.

The no-load voltage induced by this flux density in a stator winding can be calculated as

$$E_p = \sqrt{2} k_w N_s \omega_m r_s l_s \hat{B}_g \qquad [2.6]$$

where

N_s is the number of turns of the phase winding,
k_w is the winding factor (Richter, 1967; Fitzgerald *et al.*, 2003).

The main inductance of an AC machine is given by

$$L_{sm} = \frac{6\mu_0 l_s r_s (k_w N_s)^2}{p^2 g_{\text{eff}} \pi} \qquad [2.7]$$

where p is the number of pole pairs.

Slot, air-gap and end-winding leakage inductances are calculated according to Richter (1967).

2.5.2 Loss calculations

The main losses are the losses in the stator windings and the iron losses in the stator laminations. The copper losses are calculated from the currents and the resistances:

$$P_{\text{Cu}} = 3 R_s I_s^2 \qquad [2.8]$$

where the phase resistance is calculated as

$$R_s = \frac{\rho_{\text{Cu}} l_{\text{Cus}}}{A_{\text{Cus}}} \qquad [2.9]$$

where

ρ_{Cu} is the resistivity of copper,
A_{Cus} is the cross-section area of the conductor, and
l_{Cus} is the length of the conductor of the phase winding.

The length of the conductor is calculated as the number of turns multiplied by the length of a turn, where the length of a turn is estimated as twice stack length (in the slots) plus four times the pole pitch (for the end windings):

$$l_{Cus} = N_s\left(2l_s + 4\tau_p\right) \qquad [2.10]$$

The cross-section area of the conductor is the available slot area multiplied by the fill factor multiplied by the number of slots divided by the number of turns:

$$A_{Cus} = \frac{pqk_{sfil}b_{sav}h_s}{N_s} \qquad [2.11]$$

where

q is the number of slots per pole per phase,
k_{sfil} is the slot fill factor (in this example 60%),
b_{sav} is the average slot width, and
h_s is the slot height.

The specific iron losses (the iron losses per unit of mass) are calculated as the sum of the hysteresis losses and the eddy-current losses (Richter, 1967; Fitzgerald *et al.*, 2003):

$$P_{Fe} = 2P_{Fe0h}\left(\frac{f_e}{f_0}\right)\left(\frac{\hat{B}_{Fe}}{\hat{B}_0}\right)^2 + 2P_{Fe0e}\left(\frac{f_e}{f_0}\right)^2\left(\frac{\hat{B}_{Fe}}{\hat{B}_0}\right)^2 \qquad [2.12]$$

where

f_e is the frequency of the field in the iron.
P_{Fe0h} and P_{Fe0e} are the hysteresis and eddy-current loss per unit mass at given angular frequency f_0 and flux density B_0 given by the material manufacturer.

A factor of 2 is included in this equation because the flux densities do not change sinusoidally, because the flux densities are not sinusoidally distributed, and because the material properties deteriorate from the manufacturing processes, which increases the iron losses. To calculate the total iron losses, the specific iron losses in the different parts (teeth and yokes) are evaluated, multiplied by the weight of these parts and added.

2.5.3 Design optimization

Until now, this section has described how the main parameters of the machine can be written as a function of the dimensions and material properties of the machine and how the losses can be calculated. As a next step, the machine design can be optimized according to an optimization criterion. This criterion may, for example, include minimization of the generator cost and maximization of the annual energy yield as discussed by Polinder and Slootweg (2001):

$$C = C_{\mathrm{Fe}} M_{\mathrm{Fe}} + C_{\mathrm{Cu}} M_{\mathrm{Cu}} + C_{\mathrm{pm}} M_{\mathrm{pm}} - P C_E E_y \qquad [2.13]$$

where

C_{Fe} is the cost of iron (e.g. 3 €/kg),
M_{Fe} is the iron mass,
C_{Cu} is the cost of copper (e.g. 20 €/kg),
M_{Fe} is the copper mass,
C_{pm} is the cost of magnets (e.g. 50 €/kg),
M_{pm} is the PM mass,
P is a period (e.g. 5 years),
C_E is the price of a kWh (e.g. 0.1 €) and
E_y is the annual energy yield (in kWh).

This criterion means that an additional investment in the generator efficiency (to increase the annual energy yield) has to be paid back within the period P. The variables that are varied to optimize the machine design are the machine dimensions.

It is wise to use finite element programs to refine a machine design like this. The analytical equations neglect saturation, while in reality the iron circuit is mostly used in (sometimes heavy) saturation. Finite element methods can be used to check the flux density level in the teeth and to optimize the design. Furthermore, finite element methods can be used to calculate the losses more accurately, because the flux density distribution is calculated more accurately and because eddy-current losses in magnets and back-iron can be included in the calculation. These losses are mostly neglected in analytical methods.

2.5.4 Resulting design

Table 2.2 gives the main characteristics of the 3 MW 15 rpm generator design that has also been presented in Polinder *et al.* (2006). Figure 2.6 illustrates a number of characteristics. The annual energy dissipation has been calculated using a Weibull distribution for the wind speed (Polinder *et al.*,

Table 2.2 Material characteristics, main dimensions, parameters and weights of the generator

Material characteristics	
Slot filling factor, k_{sfil}	0.6
Remanent flux density of the magnets, B_{rm} (T)	1.2
Recoil permeability of the magnets, μ_{rm}	1.06
Resistivity of copper at 120° ρ_{Cu} (μΩm)	0.025
Eddy-current losses in laminations at 1.5T and 50 Hz P_{Fe0e} (W/kg)	0.5
Hysteresis losses in laminations at 1.5T and 50 Hz P_{Fe0h} (W/kg)	2
Dimensions	
Stator radius, r_s (m)	2.5
Stack length, l_s (m)	1.2
Number of pole pairs, p	80
Number of slots per pole per phase, q	1
Air gap, g (mm)	5
Stator slot width, b_{ss} (mm)	15
Stator tooth width, b_{st} (mm)	18
Stator slot height, h_{ss} (mm)	80
Stator yoke height, h_{sy} (mm)	40
Rotor yoke height, h_{ry} (mm)	40
Magnet height, l_m (mm)	15
Rotor pole width, b_p (mm)	79
Parameters	
Main inductance, L_m (mH)	4
Stator leakage inductance, $L_{s\sigma}$ (mH)	7.3
Stator resistance, R_s (mΩ)	88
Active material weight	
Iron (ton)	18.1
Copper (ton)	4.3
Permanent magnet (ton)	1.7
Total (ton)	24.1
Full load losses	
Copper losses at full load (kW)	110
Iron losses at full load (kW)	20
Total losses at full load (kW)	130

2006). This figure shows that the iron losses are rather small at full load, but at partial load, they form an important part of the losses.

2.5.5 Discussion: scaling laws and efficiency

Scaling laws

In Section 2.3, we discussed the scaling of direct drive machines based on the force density. Because we now have a known machine design, we could also extend this analysis to weight and maybe cost. The machine discussed above, with a force density of 44 kN/m^2, has a surface area of about 19 m^2 and a weight of active materials of 24 tons. If we calculate the losses, the weight and the cost of active materials per square metre, we get the results given in

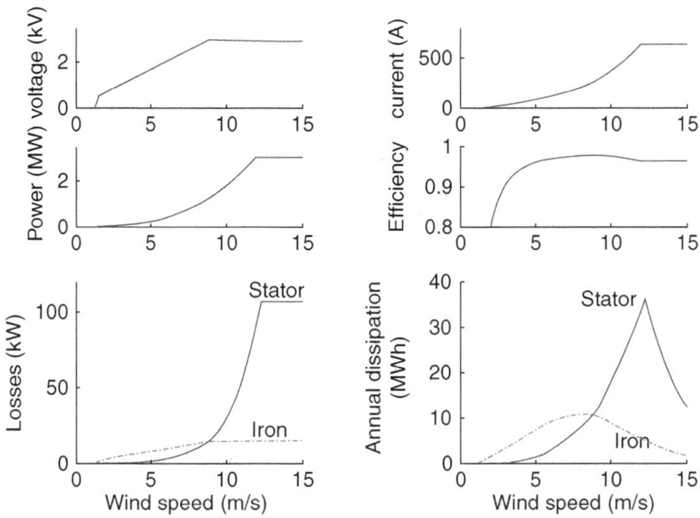

2.6 Characteristics of the generator system: line voltage, phase current, power, generator efficiency, losses and annual energy dissipation.

Table 2.3 Rough numbers characterizing active materials of direct drive generators

Maximum value of the force density (kN/m²)	44
Maximum loss density (kW/m²)	6.5
Weight of active material (kg/m²)	1300
Cost of active material in series production (k€/m²)	12

Table 2.3. It must be stressed that these numbers are very approximate and only consider electromagnetically active material (copper, iron laminations, magnets and back-iron). The indication of the cost is not valid for prototypes, but could be a rough indication for series production (assuming a cost of 3 €/kg for iron, 20 €/kg for diamond coils and 100 €/kg for magnets).

The maximum allowable force density depends on the cooling of the machine. In this example, with a force density of 44 kN/m², the losses per square metre of air-gap surface area are about 6.5 kW/m². This must be close to the losses per square metre of the Zephyros (Versteegh, 2004), so it should be possible to dissipate this with air cooling and cooling fins on the stator housing. With a water cooling system, higher force densities are possible.

Efficiency

The full load efficiency of the generator described above is about 96%. This efficiency is rather low for this power level; it is even lower than the efficiency of standard 1500 rpm 3 MW induction machines. This requires an explanation.

The efficiency of low-speed direct drive machines is physically limited. The voltage induced per unit of length of a conductor in a slot of a PM generator can mostly be calculated as

$$\vec{E} = \vec{B} \times \vec{v} \qquad [2.14]$$

where

B is the air gap flux density and
v is the relative speed between stator and translator.

It is important to realize that this expression is not always valid, and cannot be used in, for example, transverse flux machines. However, in most conventional machines (including the PM machine of Figs 2.2–2.4) it can be used. In machines with a winding in slots as described above, the air-gap flux density mostly has an amplitude close to 1 T. The speed of the magnetic flux density in the air gap is about 4 m/s. Therefore, the induced voltage in the conductor has an amplitude of about 4 V/m.

In the same conductor where this voltage is induced, there is also a resistive voltage drop if the generator is loaded. This resistive voltage drop per unit of length of the conductor can be calculated as

$$\vec{E} = \rho_{Cu} \vec{J} \qquad [2.15]$$

where

ρ_{Cu} is the resistivity of the conductor material (mostly copper) and
J is the current density in the conductor.

For values of the current density with amplitudes in the order of 5 A/mm^2, this results in a resistive voltage drop with an amplitude in the order of 0.125 V/m. If 0.125 V/m is lost in a conductor where 4 V/m is induced, it means that the efficiency is physically limited to about 97%. However, this resistive voltage drop is not only present in the slots where a voltage is induced, but also in the end windings and cable connections, where no voltage is induced. Furthermore, other losses are neglected in this explanation. Therefore, the actual efficiency is lower.

For the generator described in this section, an efficiency of 96% is very acceptable, because the efficiency of this generator system is (especially in partial load) significantly higher than the efficiency of a system consisting of gearbox and high speed generator. However, in direct drive generator systems where the speed is even lower (which is often the case in wave energy conversion), this effect is a serious limitation of the efficiency.

2.6 Future trends

Although the state-of-the-art direct drive generators are a nice solution, they have some disadvantages. The main disadvantage is that for large power levels, these generators become very large, heavy and expensive because the force density is limited and the speed is low. This is not only true for the active material, but even more for the support structure that has to keep the stator and the rotor separated by a small air gap (McDonald *et al.*, 2008; Shrestha *et al.*, 2010). A second disadvantage is that it is not so easy to make these large generators completely closed to protect them from an aggressive environment, such as offshore. A third disadvantage for low-speed direct drive applications may be that the efficiency of these machines is physically limited. Various investigations are being carried out to overcome these disadvantages.

Machines with increased force densities are investigated to reduce the size of direct drive generators. Machine types that may have higher force densities are machines like transverse flux machines, Vernier hybrid machines or variable reluctance machines. Transverse flux machines have, for example, been investigated by Dubois (2004) and Bang *et al.* (2008). Vernier hybrid machines have been investigated by Mueller and Baker (2003).

Because a significant part of the weight of direct drive generators is in the support structure, optimizing this structure is important. Stander *et al.* (2012) published a review of direct drive generator mechanical design. McDonald *et al.* (2008) worked on weight reduction of the support structure by optimizing this structure. Shrestha *et al.* (2010) investigated structural flexibility in combination with magnetic bearings as a means of reducing the weight of direct drive generators. Sopanen *et al.* (2011) worked on the mechanical dynamics of the drive train.

The main reason for the support structure becoming so heavy is that it must keep the stator and the rotor separated by a small air gap, while there are large attractive forces. In a machine with an air-gap winding or a machine without iron behind the stator coils, these forces can be reduced or even avoided. Therefore, Spooner *et al.* (2005), McDonald *et al.* (2008) and Mueller *et al.* (2009) investigated these types of machine. Making direct drive generators modular may be a way to make transportation and repair easier and cheaper. Spooner *et al.* (1996) have already worked on modular direct drive generators. Nowadays, the large generators of Enercon can be transported in parts and some other manufacturers consider the use of modules with magnets and coils to improve reparability.

The use of fractional pitch concentrated windings may lead to a significant cost reduction for direct drive generators. As indicated above, that is why they are increasingly used in PM machines. However, there may be problematical eddy-current losses in the magnets and the back-iron (Polinder *et al.*, 2007) due to the space harmonics of the winding distribution. More research is necessary to evaluate different possible combinations of numbers of poles and numbers of slots and ways of reducing the eddy-current losses in magnets and back-iron to acceptable levels.

Companies such as American Superconductor are investigating the application of superconducting coils in wind turbine generators. Abrahamsen *et al.* (2009) are investigating this further. Because superconducting coils can make much higher magnetic fields, these machines can be much smaller. The cooling system for such superconducting coils probably needs a lot of attention.

Improving availability, especially of offshore wind turbines, is an issue that needs further investigation. An important first step is the thorough analysis of failures in wind turbines, as carried out by Tavner (Spinato *et al.*, 2009). Based on this analysis, further steps have to be taken to improve availability. These steps must include manufacturing processes and protection, and repair strategies (Polinder *et al.*, 2009); they probably include condition monitoring; and they may include condition-based operation and maintenance, and fault tolerance.

2.7 References

Abrahamsen AB, Mijatovic N, Seiler E, Sorensen MP, Koch M and Norgard PB (2009), 'Design study of 10 kW superconducting generator for wind turbine applications', *IEEE Transactions on Applied Superconductivity*, vol. **19**, pp. 1678–82.

Bang DJ, Polinder H, Shrestha G and Ferreira JA (2008), 'Promising direct-drive generator system for large wind turbines', *EPE Journal*, vol. **18**, no. 3, pp. 7–13.

Cros J and Viarouge P (2002), 'Synthesis of high performance pm motors with concentrated windings', *IEEE Transactions on Energy Conversion*, vol. **17**, pp. 248–53.

Dubois MR (2004), 'Optimized permanent magnet generator topologies for direct-drive wind turbines', Ph.D. dissertation, Delft University of Technology, The Netherlands.

El-Refaie AM, Jahns TM and Novotny DW (2006), 'Analysis of surface permanent magnet machines with fractional-slot concentrated windings', *IEEE Transactions on Energy Conversion*, vol. **21**, pp. 34–43.

Fitzgerald AC, Kingsley C and Umans SD (2003), *Electric Machinery*, sixth edition, London, McGraw-Hill.

Grauers A (1996a), 'Design of direct-driven permanent-magnet generators for wind turbines', Ph.D. dissertation, Chalmers University of Technology, Göteburg, Sweden.

Grauers A (1996b), 'Efficiency of three wind energy generating systems', *IEEE Transactions on Energy Conversion*, vol. **11**, pp. 650–7.

Grauers A and Kasinathan P (2004), 'Force density limits in low-speed PM machines due to temperature and reactance', *IEEE Transactions on Energy Conversion*, vol. **19**, pp. 518–25.

Grauers A, Kasinathan P and Hamdi ES (2005), 'Force density limits in low-speed permanent magnet machines due saturation', *IEEE Transactions on Energy Conversion*, vol. **20**, pp. 37–44.

Hartkopf T, Hofmann M and Joeckel S (1997) 'Direct-drive generators for megawatt wind turbines', *Proceedings of the European Wind Energy Conference*, Dublin, pp. 668–71.

Joeckel S, Herrmann A and Rinck J (2006), 'High Energy Production plus Built-in Reliability – the New VENSYS 70/77 Gearless Wind Turbines in the 1.5 MW Class', *Proceedings of the European Wind Energy Conference*.

Lampola P (2001), 'Directly driven, low-speed permanent magnet generators for wind power applications', Ph.D. Thesis, Helsinki University of Technology, Finland.

McDonald AS, Mueller MA and Polinder H (2008), 'Structural mass in direct-drive permanent magnet electrical generators', *IET Renewable Power Generation*, vol. **2**, pp. 3–15.

Miller TJE (1989), *Brushless Permanent Magnet and Reluctance Motor Drives*, Oxford, Oxford University Press.

Mueller MA and Baker NJ (2003), 'Modelling the performance of the Vernier hybrid machine', *IEE Proceedings-Electric Power Applications*, Vol. **150**, pp. 649–54.

Mueller M and McDonald A (2009), 'A lightweight low speed permanent magnet electrical generator for direct drive wind turbines', *Wind Energy*, vol. **12**, pp. 768–80.

Polinder H and Slootweg JG (2001), 'Design optimization of a synchronous generator for a direct-drive wind turbine', *Proceedings of the European Wind Energy Conference and Exhibition*, Copenhagen, pp. 1067–70.

Polinder H, Slootweg JG, Hoeijmakers MJ and Compter JC (2003), 'Modelling a linear PM machine including magnetic saturation and end effects: Maximum force to current ratio', *IEEE Transactions on Industry Applications*, vol. **39**, pp. 1681–8.

Polinder H, Damen M and Gardner F (2004), 'Linear PM generator system for wave energy conversion in the AWS', *IEEE Transactions on Energy Conversion*, vol. **19**, pp. 583–9.

Polinder H, de Haan SWH, Slootweg JG and Dubois MR (2005a), 'Basic operation principles and electrical conversion systems of wind turbines', *EPE Journal*, vol. **15**, no. 4, pp. 43–50.

Polinder H, Damen MEC and Gardner F (2005b), 'Design, modelling and test results of the AWS PM linear generator', *European Transactions on Electrical Power*, vol. **15**, pp. 245–56.

Polinder H, vander Pijl FFA, de Vilder GJ and Tavner P (2006), 'Comparison of direct-drive and geared generator concepts for wind turbines', *IEEE Transactions on Energy Conversion*, vol. **21**, pp. 725–33.

Polinder H, Hoeijmakers MJ and Scuotto M (2007), 'Eddy-current losses in the solid back-iron of permanent-magnet machines with different concentrated fractional pitch windings', *Proceedings of the IEEE International Electrical Machines and Drives Conference*, Antalya, pp. 652–7.

Polinder H, Lendenmann H, Chin R and Arshad WM (2009), 'Fault tolerant generator systems for wind turbines', *Proceedings of the IEEE International Electric Machines and Drives Conference*, Miami, pp. 675–81.

Polinder H (2011), 'Overview of and trends in wind turbine generator systems', *Proceedings of the IEEE Power Engineering Society Annual Meeting*, Detroit.

Richter R (1967), *Elektrische Maschinen, erster Band*, third edition, Basel, Birkhäuser.

Ruuskanen V, Nerg J and Pyrhonen J (2011), 'Effect of lamination stack ends and radial cooling channels on no-load voltage and inductances of permanent-magnet synchronous machines', *IEEE Transactions on Magnetics*, vol. **47**, pp. 4643–9.

Shrestha G, Polinder H, Bang D and Ferreira JA (2010), 'Structural flexibility: A solution for weight reduction of large direct drive wind turbine generators', *IEEE Transactions on Energy Conversion*, vol. **25**, pp. 732–740.

Sopanen J, Ruuskanen V, Nerg J and Pyrhonen J (2011), 'Dynamic torque analysis of a wind turbine drive train including a direct-driven permanent magnet generator', *IEEE Transactions on Industrial Electronics*, vol. **58**, pp. 3859–67.

Spinato F, Tavner PJ, van Bussel GJW and Koutoulakos E (2009), 'Reliability of wind turbine subassemblies', *IET Proceedings, Renewable Power Generation*, vol. **3**, pp. 387–401.

Spooner E, Williamson AC and Catto G (1996), 'Modular design of permanent-magnet generators for wind turbines', *IEE Proceedings, Electric Power Applications*, vol. **143**, pp. 338–95.

Spooner E, Gordon P, Bumby JR and French CD (2005), 'Lightweight ironless-stator PM generators for direct-drive wind turbines', *IEE Proceedings – Electric Power Applications*, vol. **152**, pp. 17–26.

Stander JN, Venter G and Kamper MJ (2012), 'Review of direct-drive radial flux wind turbine generator mechanical design', *Wind Energy*, vol. **15**, pp. 459–72.

Versteegh C (2004), 'Design of the Zephyros Z72 wind turbine with emphasis on the direct drive PM generator', *Proceedings of the Nordic workshop on power and industrial electronics (NORPIE)*, Trondheim, paper number **68**.

Wu LJ, Zhu ZQ, Chen JT and Xia ZP (2010), 'An analytical model of unbalanced magnetic force in fractional-slot surface-mounted permanent magnet machines', *IEEE Transactions on Magnetics*, vol. **46**, pp. 2686–700.

3

Electrical, thermal and structural generator design and systems integration for direct drive renewable energy systems

A. McDONALD, University of Strathclyde, UK and
M. MUELLER and A. ZAVVOS,
University of Edinburgh, UK

DOI: 10.1533/9780857097491.1.51

Abstract: This chapter discusses the structural and thermal design of direct drive generators and how these generators can be in integrated into a wind turbine. The forces at play in these large electrical machines are described and the modelling approaches are discussed. The chapter introduces alternative design approaches to structural and thermal challenges in direct drive generators.

Key words: forces in direct drive generators; structural modelling of direct drive generators; integrated design of direct drive generators.

3.1 Introduction

The direct drive generator is one element of a renewable energy device. As well as being an electrical machine, it may also form an integral part of the wind turbine's structure, carrying significant loads. These direct drive generators tend to be larger and heavier than generators running at a higher speed.

3.1.1 Size of direct drive generators

The turbine blades rotate slowly (~10–20 rpm) in MW-scale wind turbines. The power of a generator is the product of the electrical machine's torque, T and its angular velocity, Ω ($P = T\Omega$) and so in order to produce high power, the directly driven electrical machine must produce a very large torque.

As a first approximation, the torque produced by a conventional rotating electrical machine can be modelled by a cylinder and a shear stress on the surface of the cylinder (Fig. 3.1):

$$T = 2\pi\sigma R^2 l \quad [3.1]$$

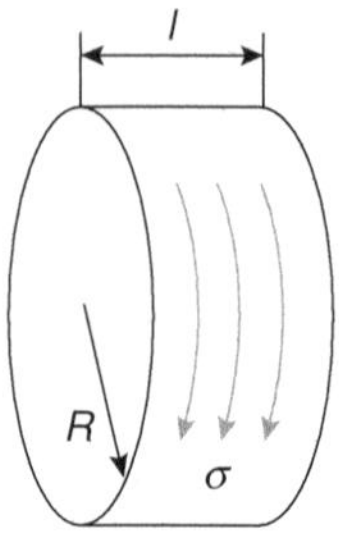

3.1 Cylinder model of torque produced by a generator.

where R is the airgap radius, l is the axial length of the generator and σ is the electromagnetic shear stress.

There are practical limits to electrical and magnetic loading in electrical machines and thus there is an upper limit to the shear stress that can be developed. Designers typically use shear stress of the range σ = 25–50 kN/m^2.[1] The implication of this ceiling is that the radius and length of the directly driven generator become very large when compared to higher-speed machines at a similar power level.

If a generator designed to run at, say, 1500 rpm is connected to rotor blades rotating at 15 rpm (with a gearbox of 1:100 step-up ratio) then the generator only has to produce 1% of the torque of the equivalent direct drive electrical machine. If the gearbox-driven and directly driven generators are of the same generator topology (e.g. they are both radial-flux permanent magnet synchronous generators which develop the same shear stress, σ in the airgap) but have different dimensions R and l (where the aspect ratio $l/2R$ = constant), then the direct drive electrical machine will have a radius and length approximately 4.5 times greater than that of the original geared electrical machine. Direct drive leads to generators which are bulky relative to their electrical power output.

3.1.2 Forces in electrical machines

From a design viewpoint, maintaining the airgap clearance between the rotor and stator is critical. While the forces described in this section are common to all electrical machines, it is in large electrical machines that they become structurally demanding. Larger machines tend to have larger forces – due to larger surface areas – which act at larger radii or distances from the nacelle mounting points.

A typical design has an airgap clearance of 0.1% of the airgap diameter,[2] meaning that a generator with an airgap diameter of 5.0 m will have an

3.2 Direct drive permanent magnet generator.

airgap clearance of 5 mm. The maximum allowable deflection should only be, say, 10–20% of this clearance, otherwise the airgap flux density might vary significantly[3–5] and hence affect electromagnetic forces and possibly add to the magnitude of some forces.

Figure 3.2 shows a radial-flux permanent magnet generator. The inner rotor is essentially a cylinder with permanent magnets mounted on the outside of the cylinder. Only part of the stator is shown for simplicity. A rectangular window is shown for enlargement purposes in Fig. 3.3. Some of the forces at play in the generator include:

- **Shear stress/torque transmission**: The area near to the airgap is where mechanical energy is converted into electrical energy (Fig. 3.3, top). For the generator, a rotor transmits mechanical torque to the airgap and the stator must be able resist that same torque.
- **Normal stress**: The magnets on the rotor lead to a large force of attraction between rotating and stationery parts of the generator (Fig. 3.3, middle).
- **Wind turbine loading**: Depending on the integration and location of the direct drive generator into the wind turbine and drivetrain it may see

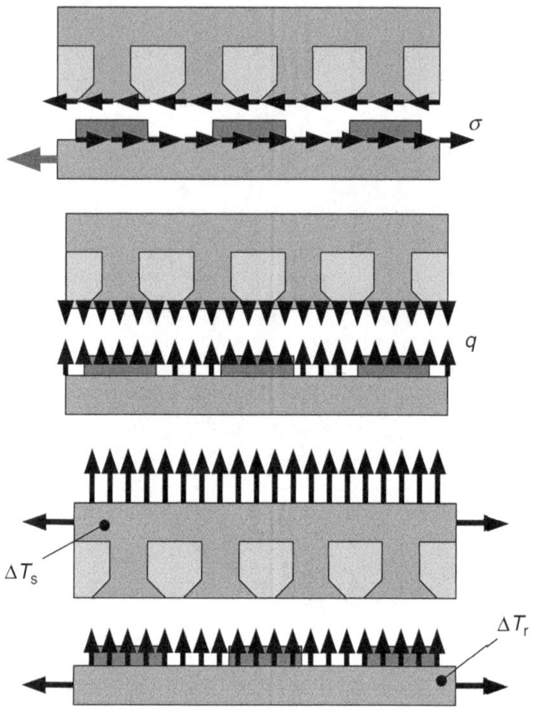

3.3 Forces in direct drive generators: shear (top), normal (middle) and thermally induced (bottom).

some loading from the rotor blades, weight of the rotor blades, horizontal and vertical wind shear, yaw error and inertial effects.

- **Gravity**: During assembly, transportation and installation (particularly when the axis is off the horizontal) the weight of the generator itself can be an important load to consider.
- **Thermal strains**: Significant amounts of heat are generated in electrical machines leading to temperature rises in machine parts. Different temperature rises, falls and rates of change can give rise to differential thermal expansion and contraction (Fig. 3.3, bottom).
- **Centripetal forces**: At low speeds this is a relatively small load.

3.1.3 Electromagnetic, thermal and structural design

Wind turbine manufacturers often wish to constrain or reduce the large size and weight of direct drive generators. In order for the generator design to be successful, the connections between electromagnetic, thermal and structural design elements ought to be well understood. Figure 3.4 gives some example relationships between these design elements.

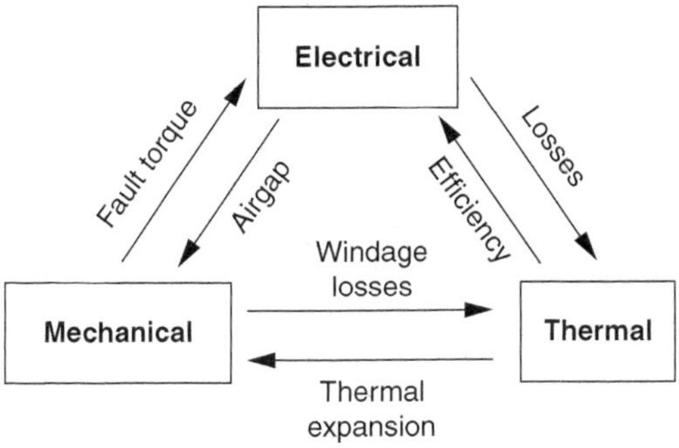

3.4 Example interactions between electrical, thermal and mechanical design aspects of the generator.

3.2 Integrated systems design of machine topologies

This section will explore different approaches to designing direct drive generators.

3.2.1 Design perspectives

Because of their complexity, size and cost, direct drive generators require the attention of several disciplines during design and manufacture. When it comes to designing direct drive generators, different engineering disciplines have quite different priorities. A good design in terms of cost, performance and reliability will be able to take all these perspectives into consideration and reach a suitable compromise.

To illustrate the different perspectives, Figs. 3.5–3.8 show the priorities with which an electrical engineer, a mechanical engineer, a heat transfer specialist and a manufacturing engineer would view the generator.

An electrical engineer (Fig. 3.5) hopes to get maximum performance (e.g. power output, efficiency) using the smallest possible amounts of material such as copper and permanent magnet. This may lead to a small physical clearance between the rotor and stator and a large radius. Choice of material and its form is important – such as high fill factors of copper and design of steel laminations. Electrical insulation is also a top priority for the electrical engineer.

The electrical design directly accounts for around 52% of the cost of a 3 MW direct drive permanent magnet generator.[6]

The mechanical design engineer (Fig. 3.6) is more interested in the physical robustness of the generator, in terms of strain, stress, strength and fatigue. Elegant mechanical design can produce suitable safety factors using low-cost and lightweight assemblies. The pursuit of robustness might mean choices of materials and dimensions which are electromagnetically non-optimal. Some electromagnetic parameters can be optimised to reduce items such as torque ripple and fault torques, thereby helping the mechanical design.

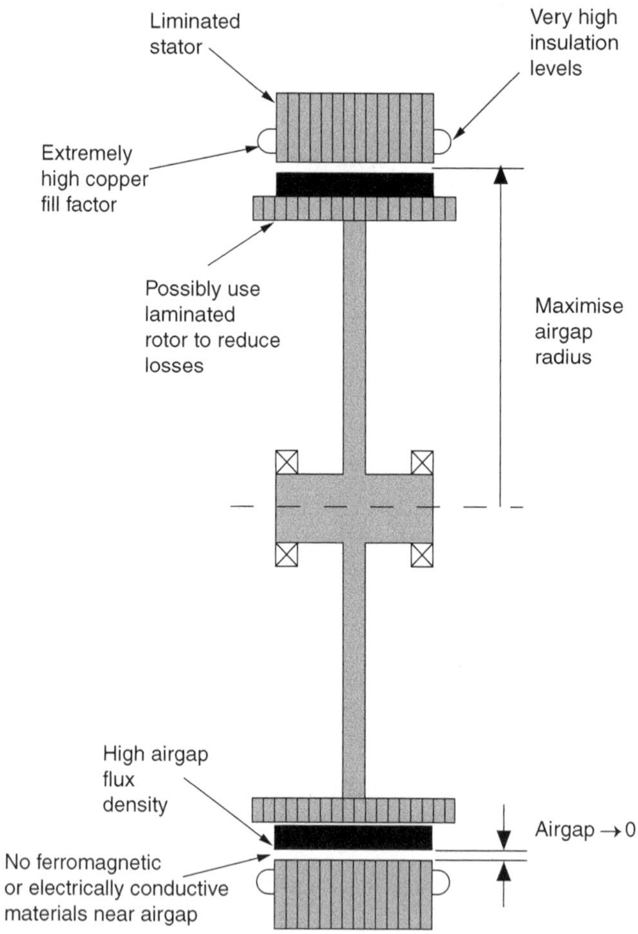

3.5 Electrical design perspective.

The mechanical design directly accounts for around 55% of the mass of a 5 MW direct drive permanent magnet generator.[7]

The cooling of the generator (Fig. 3.7) has a different set of requirements. The heat transfer specialist will focus on adding to and maximising thermal conduction and convection paths so as to lower winding temperatures.

The performance of the following electrical materials is temperature dependent:

- The resistance of copper is proportional to temperature, so I^2R losses are proportional to winding temperature too. Reducing temperature can lead to increased power density and higher efficiencies.

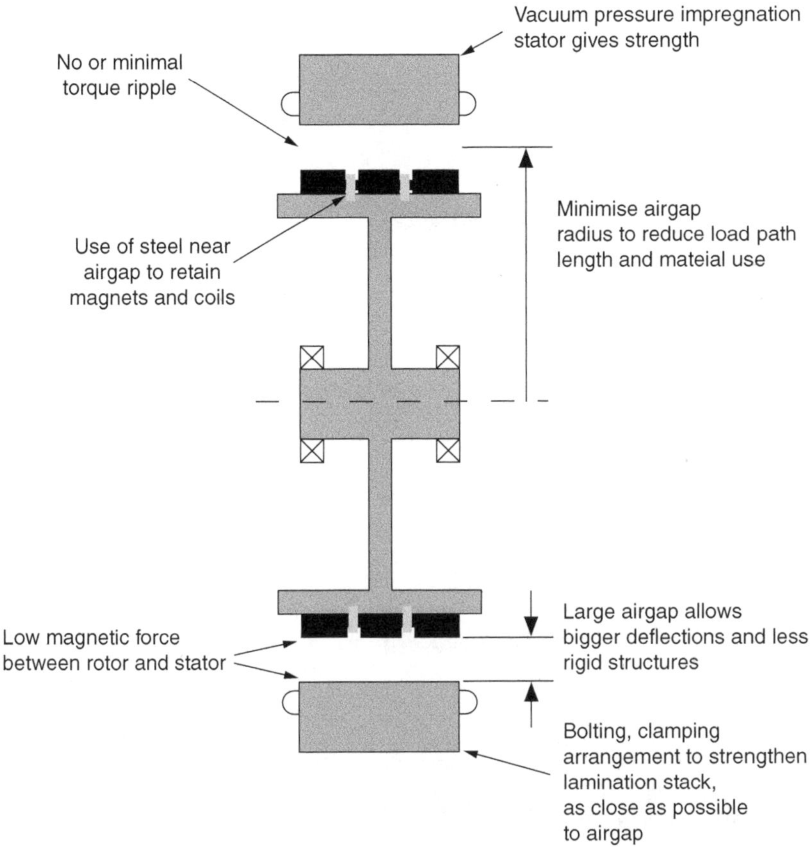

3.6 Mechanical design perspective.

- The magnet BH curve is temperature-dependent, so higher temperatures lead to reduced magnetic loading and increased risk of demagnetization in fault cases.

From a manufacturing engineering perspective (Fig. 3.8), simplification of the design can lead to reduced manufacturing and assembly costs. Modularisation and less strict tolerances can also lead to cheaper generators.

3.2.2 Flowchart of design process

The early stages of the design process are usually dominated by the electrical design, with other aspects following on from that (Fig. 3.9). This can mean that

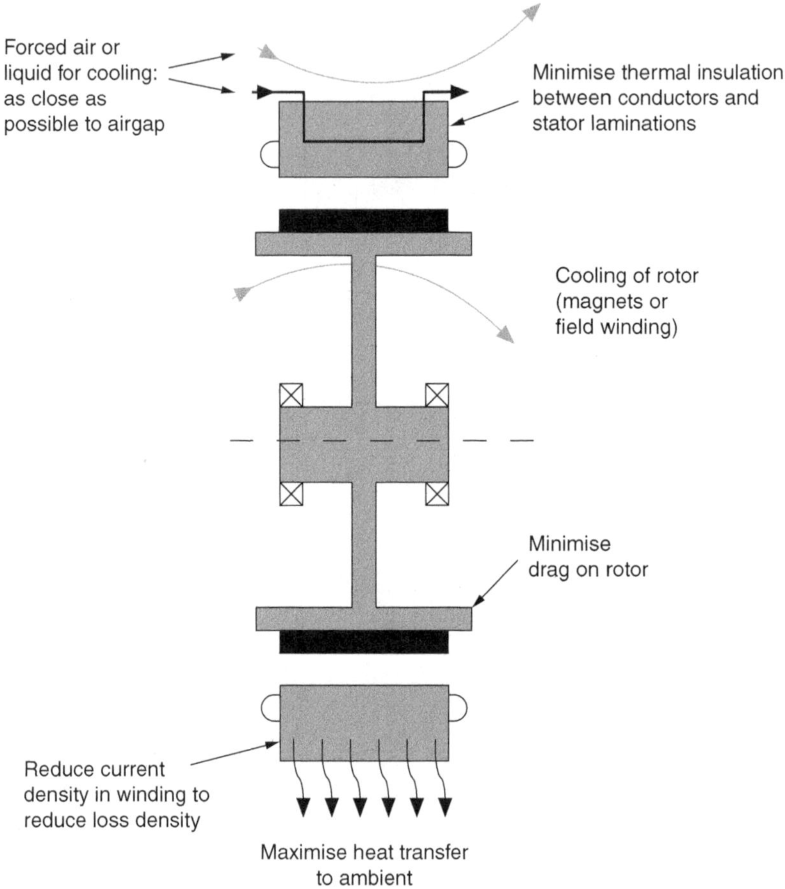

3.7 Thermal design perspective.

key dimensions, forces and thermal parameters are assumed and that mechanical and thermal designers are constrained. It may be an electrical optimum (in terms of cost, mass and efficiency) but not perhaps a global optimum.

Figure 3.10 shows an alternative, integrated design approach whereby the mechanical and thermal design is considered at the same time as the electrical design.

3.3 Structural considerations and mechanical design

The direct drive generator must deal with a number of structural challenges because it is a large part of the wind turbine and also because there are significant forces at play within the generator.

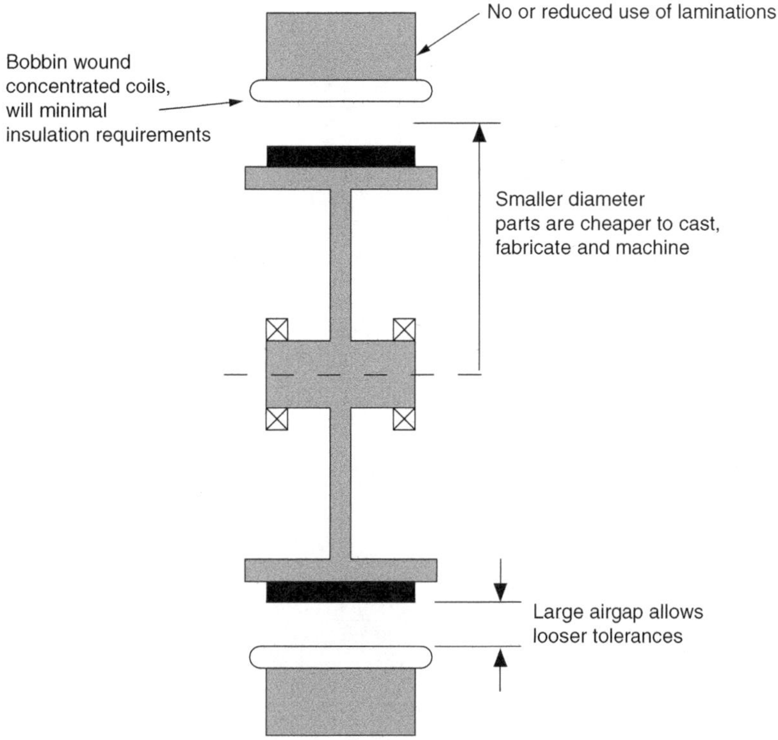

3.8 Manufacturing design perspective.

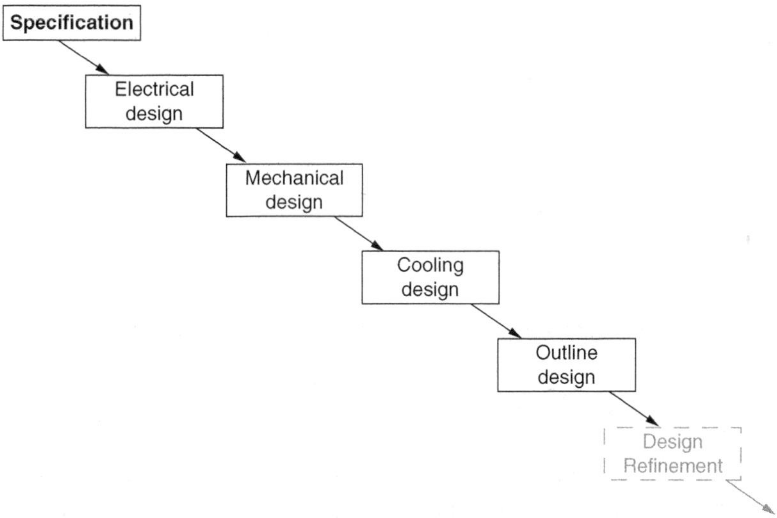

3.9 Early design stages – traditional approach.

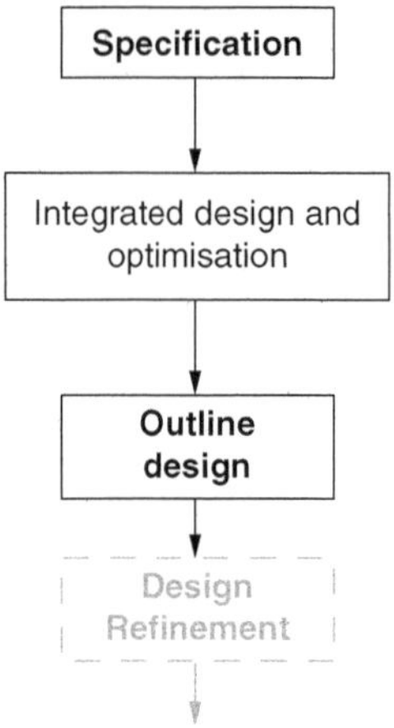

3.10 Early design stages – integrated approach.

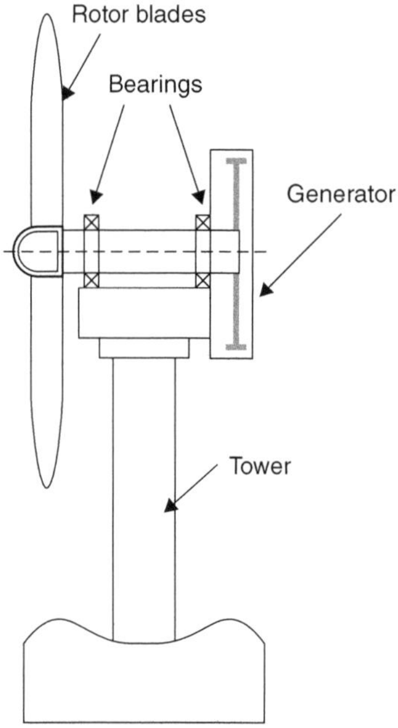

3.11 Generator downwind of the tower, for example, Scanwind/GE.[8]

3.3.1 Direct drive wind turbine layout

There are a number of different ways that a direct drive generator can be mounted onto the turbine. The first main variation is the location of the direct drive generator relative to the tower. Figures 3.11–3.13 diagrammatically

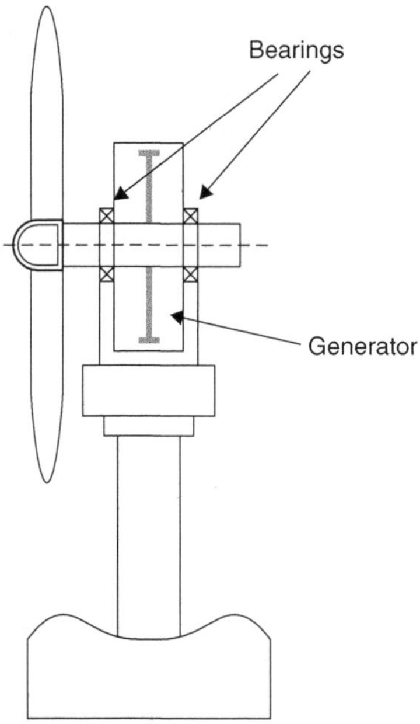

3.12 Generator over the tower, for example, MTorres.[9]

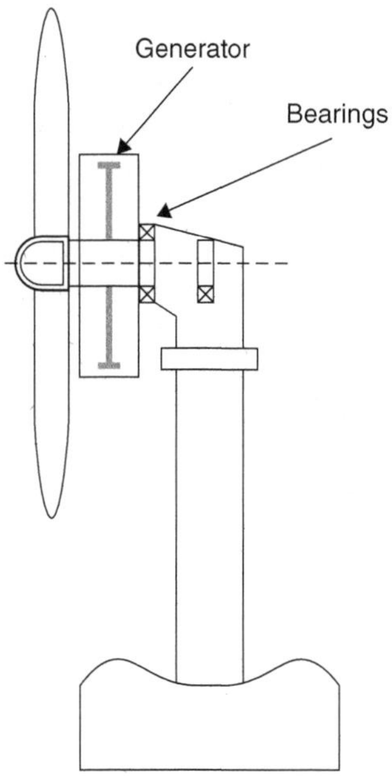

3.13 Generator upwind of the tower with a single bearing, for example, STX93.[10]

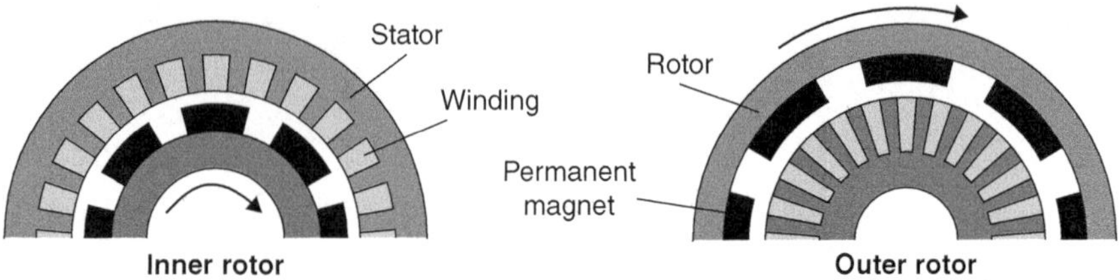

3.14 Inner and outer rotor generator variants.

show a generator downwind of a 4 MW GE wind turbine, an MTorres 1.5 MW wind turbine with the generator mounted on top of the tower and a 2 MW STX wind turbine with generator mounted upwind of the turbine tower. The upwind and downwind options allow the generator (with its large diameter) to hang clear of the turbine, and the bearings can sit low down in the nacelle structure. The upwind option is popular because it does not require a long shaft and allows a single bearing arrangement, which in some cases leads to a reduction in cost. One of the downsides is that the generator becomes a structural load path of the turbine itself and it is difficult to remove the generator without having to take out the turbine rotor as well.

Some of these options were studied in the Northern Power Systems WindPACT Drive Train Alternative Design Study Report.[11] A comparison based on cost, weight, risk, shipping, assembly and serviceability led to the single bearing option being chosen – this is a single main bearing mounted on a spindle which also carries the stator. This assembly is connected to a 'turret' which carries the turbine rotor loads down to the tower. One of the main downsides of this arrangement is that the turbine rotor is offset from the tower and so the tower and other items in the load path must be stiffer and thus typically they are heavier.

A further variation in direct drive generators is the difference between having an inner or an outer rotor (Fig. 3.14). The inner rotor means that the stator is easier to cool naturally; an outer rotor generator may be more compact – the magnets and rotor yoke tend to have a shallower profile than the stator teeth and yoke – and the centripetal force on the magnets aids their seating onto the rotor.

3.3.2 Challenges – forces

The **shear stress** is the useful force in an electrical machine, giving rise to torque as shown in Equation [3.1]. Designers have long tried to maximise this shear stress; hence the interest in transverse-flux permanent magnet machines which are characterised by high shear stresses.[11] If the distributions of flux density and electric loading are sinusoidal and are displaced by δ and the peak values are $\hat{B}$ and $\hat{K}$ then[12]

$$\sigma = \frac{1}{2}\hat{B}\hat{K}\cos\delta \qquad [3.2]$$

Because the shear stress is perpendicular to the airgap – as shown in Fig. 3.3 – it does not serve to close the airgap in symmetrical structures. Sometimes generators are asymmetrically supported – as shown in Fig. 3.13 – and so deflections into the airgap can occur.

During fault conditions this shear stress can be many times greater than the nominal value. For example, in Reference 13 for a three-phase short circuit the peak fault current is ten times rated current and the steady-state fault current is twice rated current.[13]

The normal component of Maxwell stress, q is directed across the airgap, so that the outer iron surface (normally the stator) is attracted radially inwards and the inner iron surface (usually the rotor) is attracted radially outwards. This is shown for a section of the radial-flux machine in Fig. 3.3. When a large airgap flux density is used (e.g. $\hat{B} > 0.8$ T) this stress can be greater than 255 kPa, or an order of magnitude greater than that of the shear stress. The normal component of Maxwell stress is a function of the square of the airgap flux density,

$$q = \frac{\hat{B}^2}{2\mu_o} \qquad [3.3]$$

where μ_o is the permeability of free space. In the rest of this chapter, the normal component of Maxwell stress will be referred to as the normal stress.

Generator weight

The tilt angle, ψ of a horizontal axis wind turbine rotor axis to the horizontal (typically about 5° in a conventional wind turbine) means that gravity acts on the generator along two axes: there is a major and a minor component.

The major component of **gravity** is of magnitude $g\cos\psi$ and acts to deflect the rotor and stator back iron and structural members (such as spokes and support spiders). If the support is one-sided then there will be a tipping moment. Depending on their construction and the distribution of active material in the rotor and stator there may be different deflections, causing a narrowing or widening of the airgap. By way of comparison, if the stator cylinder was 15 tonnes and 1 m long, with radius 2.5 m, then the force per unit area would be about 8 kPa.

The minor component of gravity is $g\sin\psi$ and is axially directed leading to misalignment of the rotor and stator in the axial direction. In a vertical axis wind turbine this becomes a more significant challenge.

Thermal strain

Significant amounts of heat are generated in electrical machines leading to temperature rises in machine parts. These in turn give rise to expansion given by

$$\Delta L = L_o \alpha \Delta T \tag{3.4}$$

where ΔL is the change in dimension, L_o is the original dimension, α is the thermal coefficient of expansion of the material and ΔT is the temperature rise. A difference in temperature rise in the rotor and the stator of a radial-flux machine will lead to a change in the airgap clearance as shown in Fig. 3.3. An aluminium outer rotor ($\alpha = 22 \times 10^{-6}\ K^{-1}$) with a temperature rise of 40 K with a steel inner stator ($\alpha = 13 \times 10^{-6}\ K^{-1}$) with a temperature rise of 80 K (class B) would see the airgap opening up by 3 mm.

3.3.3 Modelling approaches

In order to employ an integrated design approach – such as that in Fig. 3.10 – one needs to be able to link electromagnetic and structural models. A number of approaches can be used at different stages of the design process.

Cost modelling through scaling

Grauers[2] proposed a scaling approach which characterises the cost of a generator based on generator diameter and length, relative to dimensions of reference structures (Fig. 3.15). The cost of a generator structure is given as

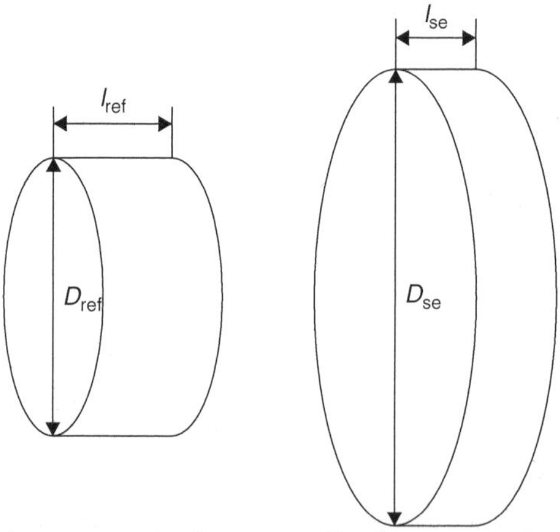

3.15 Grauers' structure cost modelling through scaling.

$$C_{\text{str}} = c_{\text{str}} \frac{1}{2}\left[\left(\frac{d_{\text{se}}}{d_{\text{ref}}}\right)^a + \left(\frac{l_{\text{tot}}}{l_{\text{ref}}}\right)^a\right] \quad [3.5]$$

where d_{se} is the stator outer diameter, l_{tot} is the axial length, d_{ref} and l_{ref} are the outer diameter and axial length of a reference stator and c_{str} is the specific cost of the structure.

The exponent $a = 3$ was estimated on the basis that a structure of 2 m diameter and 1 m length is approximately 20 k€ and a similar structure but for a smaller generator of 1 m diameter and 0.5 m length is estimated to be 2.5 k€.

This can be a useful approach when varying or optimising a design about a particular design point. It assumes a certain manufacturing approach with a constant specific cost, whereas different sized structures often require different manufacturing methods either due to cost or size limitations; different manufacturing approaches tend to have different specific costs. It also assumes that the cost of the structure scales proportionally with the dimensions of diameter and length without reference to the loads that the structure is designed to deal with.

Analytical approaches to strain modelling

The primary function of the generator structure is to maintain the clearance (airgap) between rotor and stator. This means that strain, rather than stress, is the primary focus of attention.

One analytical approach has been suggested by Tavner and Spooner.[14] They identified a number of ways that localised contact between the rotor and stator can occur:

1. Radial expansion of the rotor or radial compression of the stator.
2. Radial relative displacement.
3. Distortion of either or both of the circular surfaces into ellipses (known as ovalising).
4. Distortion with ripples around the circumference.

These can be described by:

$$\delta = \hat{\delta} \sin n(\theta - \varphi) \quad [3.6]$$

where θ is the angular position and ϕ is the orientation of the deformation pattern and n is an integer describing the wave number of the pattern. Figure 3.16 shows the first modes of deflection, $n = 0$ for deformation of type a above, $n = 1$ for type b, $n = 2$ for type c and $n \geq 3$ for type d.

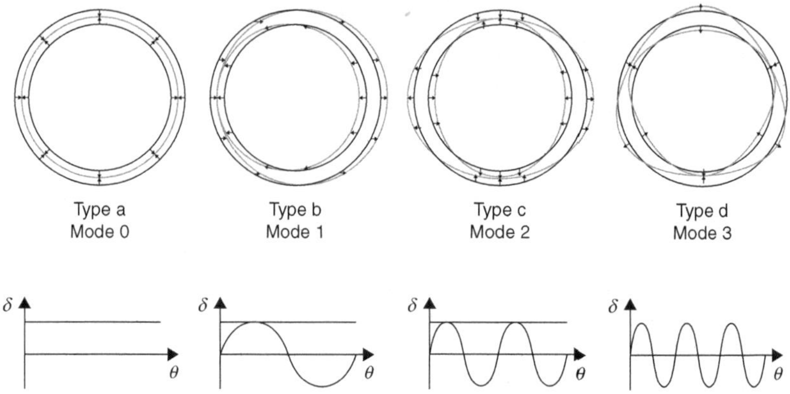

3.16 Deflection modes for a radial-flux machine.

By looking at the magnetic-elastic instability Tavner and Spooner found that the minimum stiffness is in mode 1 and they produced a limit relating structural stiffness to the maximum torque possible.

$$T_{\max} = 2\pi k_{s,\min} p g^2 \quad [3.7]$$

This stiffness can be linked back to dimensions and mass using Equation [3.8].

$$k_{s,\min} = \frac{EA_{\text{arm}}}{4(R-r)} \quad [3.8]$$

Analytical approaches: radial-flux machines

In radial-flux machines, McDonald assumed simple disc and arm structural models to link the electromagnetic and mechanical design.[15] These modelled radial, axial and tangential deflections (Fig. 3.17).

Analytical approaches: axial-flux machines

For an axial-flux machine the structural deflection can be linked to magnetic forces and structural dimensions using simple disc models,[16] Fig. 3.18.

Numerical approaches

A more advanced approach to modelling the structural behaviour of the generator is to use computational finite element methods, such as ANSYS.[17,18] These are now standard design tools and the results can be used with a high level of confidence. Their versatility allows the designer to model

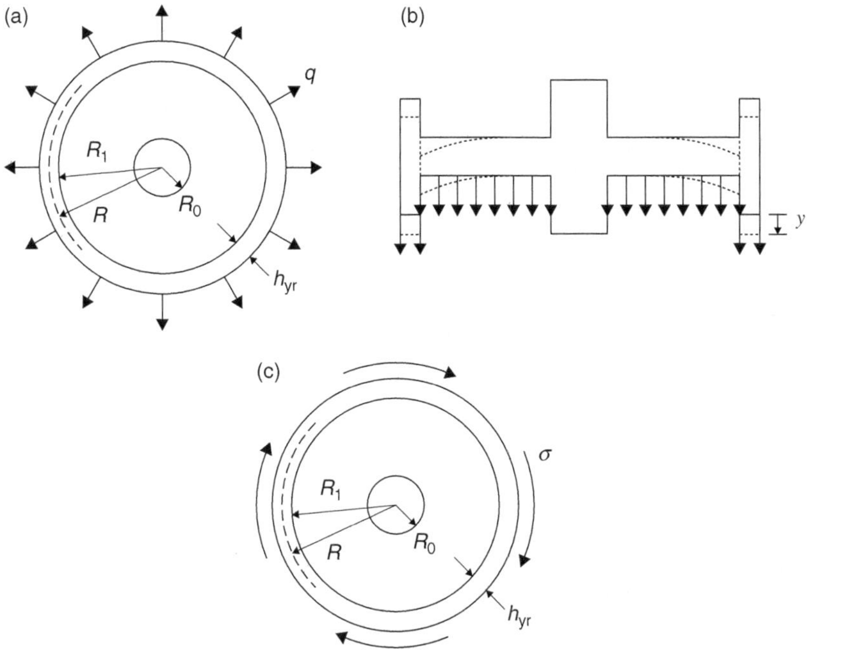

3.17 Radial-flux models: radial (a), axial (b) and tangential deflection (c).[15]

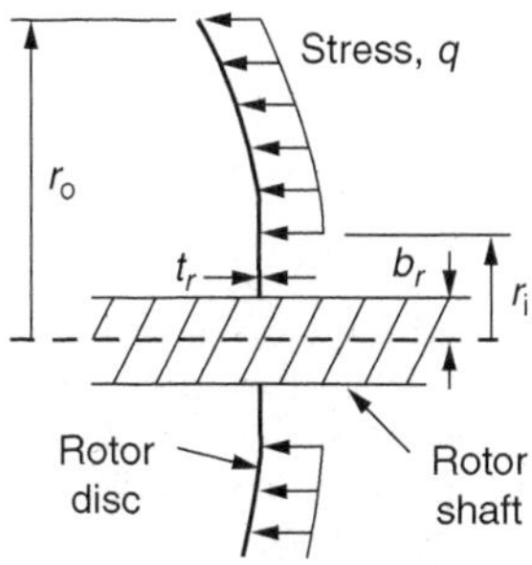

3.18 Axial-flux model.[16]

dynamic behaviour (such as response to vibrations) and understand fatigue properties.

Unfortunately they are still computationally intensive so their uses in the very early stages of design and in global generator optimisations are limited.

Optimisation

The analytical models can be linked to the electromagnetic model. So for example, Equations [3.2] and [3.3] give the tangential and normal loading.

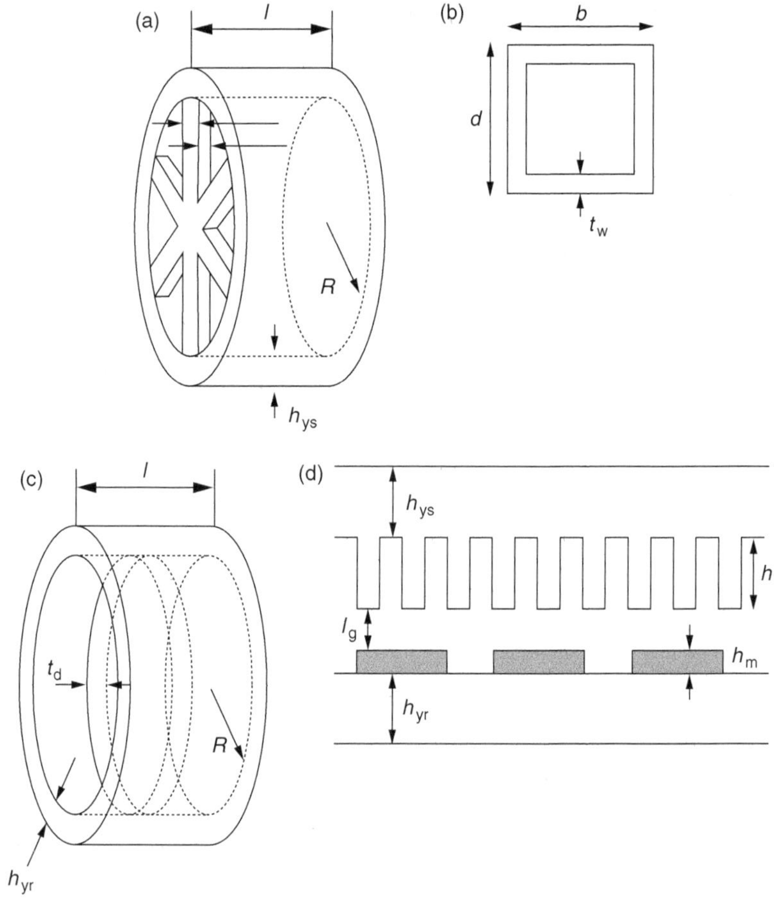

3.19 Depiction of the variable dimensions that were used for the optimisation of the generator structures. (a) The variables that describe a structure with arms. (b) The variables that describe the structure's hollow arms. (c) The variables that describe a rotor with disc. (d) The variables that describe the electromagnetic model.

It is then possible to use the models for optimisation purposes. McDonald *et al.* used a genetic algorithm (GA) toolbox to minimise the structural mass of a radial-flux PM direct drive rotor.[19,20] Zavvos *et al.* expanded McDonald's original optimisation tool in order to include both the active and structural design of the rotor and the stator of a Permanent Magnet Direct Drive generator structure.[21] The most common cases of such generator structures were modelled and optimised. The structural variables that could describe a Permanent Magnet Direct Drive structure with discs or arms were free to vary in an attempt to find the best combination of parameters that would give the most lightweight or cheap structure (Fig. 3.19).

For cases where the active material should be optimised as well, the machine's airgap (l_g) and magnetic flux (B_g) were included as variables in the process.

The best combination of parameters is achieved through a process that imitates evolution. The given parameters are randomly distributed through the multi-dimensional design space and candidates are changed in order to create 'generations' of the object under examination. Each generation is evaluated according to the given objective function that returns a rating for each one of them. At the end, the parameters of the 'best' generation – meaning the one that scored the highest – are returned to the user. This process can easily be adapted to optimise the weight or the total cost of the generator, depending on the objective function. The objective function that rates the random structures is based on the resulting weight or cost of the generator and the calculated deflection in the airgap.

These models were modified by Zavvos *et al.* to fit two different Transverse Flux Permanent Magnet Direct Drive topologies.[22] The two Transverse Flux Permanent Magnet Direct Drive machines are explicitly described by Bang in Reference 23. The main difference between these two Transverse Flux topologies is that the airgap is radially oriented for Transverse Flux Permanent Magnet Direct Drive 1, while axially oriented for Transverse Flux Permanent Magnet Direct Drive 2. Different analytical models to calculate the deflection due to the attraction forces were created in each case, while the models calculating the deflections due to gravitational pull and tangential force remained the same as the Radial Flux Permanent Magnet Direct Drive case.

The results of this analytical approach showed that higher aspect ratios could lead to lighter and cheaper generator structures for radial-flux topologies, while small aspect ratios should be preferred for transverse-flux topologies. The comparison of the total mass of all the resulting lightweight structures showed that transverse-flux machines had the best mass/torque ratio, but only if the generator's topology was that of the Transverse Flux Permanent Magnet Direct Drive.[22]

3.3.4 Design approaches and alternative solutions

The previous section showed how the airgap between rotor and stator can be maintained by making the rotor and stator structures stiff. It has been shown that this leads to heavy generators when the torque rating becomes significant and that the generator weight grows at a greater rate than the turbine power rating. An alternative approach to maintaining the airgap clearance is to shorten the load path by using a bearing (or bearings) local to the airgap. Figure 3.20a shows the NewGen concept where two sets of wheels on the stator roll on a machined surface on the rotor.[24] This means that the main load path reduces from, say, 5–10 m to perhaps less than 1 m. Much of the generator structure can be designed according to torque transmission rather than normal stress as the load, leading to reduced structural weight. NewGen claim 50–80% less material than in a conventional generator without this bearing arrangement. Figure 3.20b shows a hydrostatic bearing concept[7] and Fig. 3.20c shows a magnetic bearing concept that performs the same functions.[25] Each embodiment has different merits and drawbacks in terms of cost of the bearing systems, weight, reliability, operation and maintenance costs, active control requirements, energy use and access to generator components.

These approaches have not yet been successfully adopted, perhaps because they introduce extra components with additional failure modes which could be seen to reduce the often cited reliability advantage of direct drive over geared drivetrain.

As the normal force (Equation [3.3]) becomes more dominant, an alternative solution is to redesign the magnetic circuit. Removing iron from the stator produces a so-called 'ironless' or 'air cored' machine and there is no normal force of attraction between rotor and stator and so the structure can be less stiff and lighter. As the magnetic loading is less, shear stress is reduced (Equation [3.2]) and so the generator can become very large, for example Fig. 3.21.[12]

3.4 Thermal considerations

All electrical machines generate losses which manifest themselves as heat. The material properties of magnets and insulation materials can also limit the peak temperatures allowed in a generator. The rise in temperature is effectively a limit to electrical loading and thus torque. Different cooling arrangements allow markedly different electrical loadings: reference 26 gives 30–80 kA/m for air cooling and 150–200 kA/m for direct water cooling.

Previous sections showed some of the links between thermal management and electrical and structural performance. There are some important differences between thermal management in low-speed and higher-speed generators used in wind turbines:

- **Size of the generator:** For a given power rating, the direct drive generator will be much larger. This means that the surface area available for heat dissipation is bigger, so passive heat transfer is better. Because of their size it does mean however that there is less (radial) space for water jackets and other similar auxiliaries if they are needed.

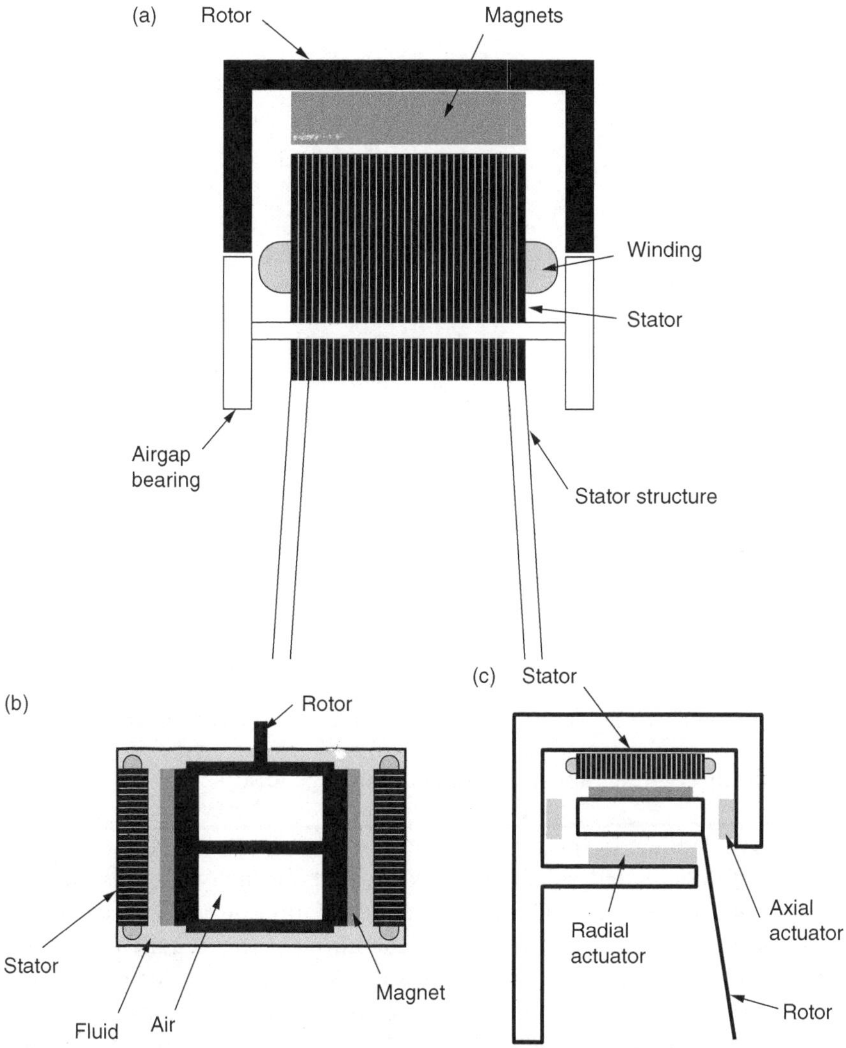

3.20 Shortening of the load path between rotor and stator using various airgap bearing technologies. (a) Steel wheels,[24] (b) hydrostatic bearing[7] and (c) magnetic bearing.[25]

- **Location of the generator:** A high speed generator will be located within a nacelle. In order to transfer heat away a heat exchanger needs to be employed linking to a radiator mounted on top or at the back of the wind turbine nacelle. The outer skin of a direct drive generator will often be in direct contact with the (moving) outside air.

3.21 Ironless radial-flux generator.[12]

3.4.1 Cooling approaches in direct drive wind turbines

There are three broad approaches to cooling a generator:

1. **Passive cooling**: In this arrangement the heat produced within the generator is conducted, radiated and convected internally towards the outside of the generator. Convection and radiation on the outside of the generator then transfer the heat to the air flow. This is a nice approach as it does not depend on auxiliary systems such as fans and pumps which require servicing from time to time. Some elegant solutions have been found to take advantage of the wind air flow (moving air gives better heat transfer than when there is no wind), for example in the Vensys 1.5 MW shown in Fig. 3.22.[27]
2. **Forced air cooling:** In this arrangement fans are used to circulate air within the generator. This aids heat transfer as convection coefficients are improved with air velocities. Although this takes up power (to run the fans) the reduction in winding temperatures and the resulting efficiency and improved power density gains are worth the extra cost and energy consumption. As the nacelle is generally sealed from the outside, a heat exchanger system is required transferring heat from an internal circuit to the wind, for example in the Vensys 2.5 MW turbine (Fig. 3.23).[28]

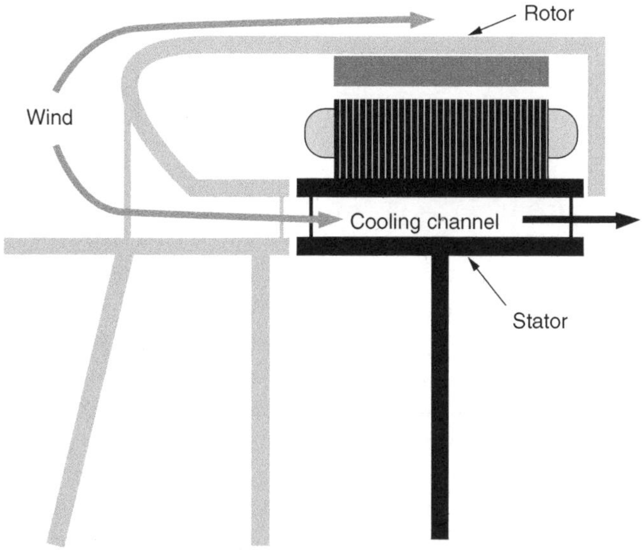

3.22 The wind is used to cool the back sides of both the rotor and stator (through internal ducts) of the VENSYS 1.5 MW wind turbine.

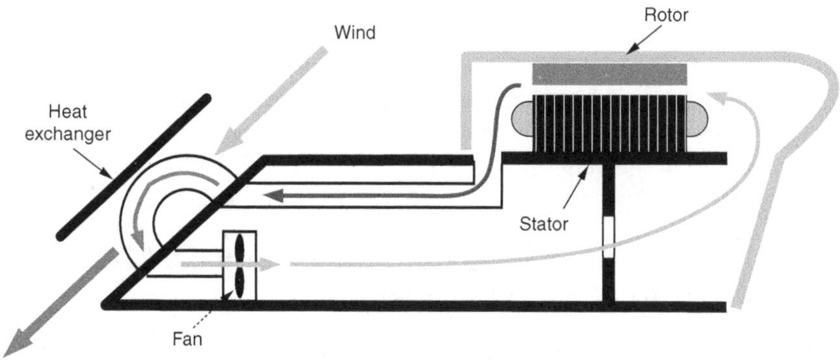

3.23 The Vensys2.5 MW uses fans to cool the generator through a heat exchanger.

3. **Liquid cooling:** Eventually there is a limit to the amount of heat transfer that forced air cooling can achieve, either through fan energy consumption or ducting cross-sectional area. Liquid cooling can be used to transfer the heat from the generator to a radiator. By moving from air cooling to liquid cooling, Enercon have been able to increase power rating of the E-82 from 2.0 to 3.0 MW while using a 5.5 m diameter generator.[29] Reference 30 suggests that a liquid-cooled generator can be 60% of the diameter and mass of an air-cooled generator.

3.4.2 Thermal modelling of electrical machines

The starting point for the generator cooling designer is often the temperature classifications:[31]

B 130°C
F 155°C
H 180°C

which may be defined in the design specification.

Within each class there can still be a significant impact on efficiency. For example, as copper has a temperature coefficient of resistivity of $\approx 4 \times 10^{-3}$/K, a 25 K temperature rise would lift I^2R losses by about 10%. In order to model the temperature rise in the generator, lumped parameter models can be used quickly and effectively, for example.[32,33] These models use an electrical circuit analogy where temperature difference corresponds to voltages, heat flux is current and conductance can be used to model thermal conductance, convection and radiation. Grauers used this method to model temperatures in a direct drive 600 kW permanent magnet generator.[2] The main drawback is that there are limited nodes, so variation within generator elements can be difficult to model accurately. For more in-depth analysis there are a number of finite element analysis tools which will model heat transfer; these have the potential for greater accuracy but are slower. For either of the approaches, convection heat transfer coefficients are important. Often computational fluid dynamic software will be used to estimate airflow and convection within the generators.

3.5 Designs of machine topologies for 5–20 MW direct drive wind turbines

Evolution in wind turbines creates the demand for larger power outputs and increased reliability. As wind turbines increase in power output, their size and mass grows as well. Scaling techniques are used in an attempt to calculate the expected structural or electromagnetic parameters of a larger generator based on data from similar but smaller existing generators.

Direct drive generators operate at low speed. When scaling up in power P, the rotational speed of the wind turbine ω_m decreases. This is because there are rotor blade tip speed upper limits, so larger turbines with larger blades rotate more slowly. Increasing power and reduced rotational speed both lead to higher torque requirements. Equation [3.1] showed how torque is related to generator diameter and axial length.

McDonald outlined that the structural mass of a permanent magnet direct drive (PMDD) generator with power output of 5 MW or greater can reach

80% of its total weight.[15] A number of machine topologies proposed in literature were compared by Bang[23] for their total mass when scaling from 5 MW up to 20 MW. The total mass of three proposed radial flux (RF) PMDD topologies[3,10,24] was compared against the mass of a commercial electrically excited direct drive generator from Enercon[29] and the conventional doubly-fed induction generator with a three-stage gearbox (DFIG 3G). The total mass/torque ratio of generator concept was kept constant when scaling up their power output.

The results showed a large variation between the mass of the different concepts and designs: the heaviest was 3.5 times the mass of the lightest (600 tonnes for 20 MW). What is clear is that all of the current solutions are heavy and that scaling existing technology without modification would lead to high top head mass in the generator and related high costs for lifting and assembly.

3.6 Application to direct drive marine energy systems

Most of the discussion so far in this chapter has been related to wind turbines. Many of the same points are applicable to tidal and wave power. Some of the challenges are greater for marine energy. For example the speed input into a tidal turbine will be less than that for an equivalent rated wind turbine, so the direct drive generator will tend to be larger and more expensive. The marine environment means that corrosion protection and sealing are more demanding for tidal and wave power generators than for wind.

3.6.1 Direct drive tidal power

One of the best examples of a direct drive generator in a tidal turbine is that of Open Hydro.[33] Here the direct drive generator has a rotor which is integrated onto the outside of the open centre turbine rotor (Fig. 3.24). The magnets are mounted onto the rim of the rotor. This is structurally elegant as the torque path from the blades to the generator airgap is short, the generator and the turbine are closely integrated and the generator airgap radius is maximised. From Equation [3.1] it is clear that this can lead to short axial length and a reduced requirement for shear stress. The fact that the stator sits on the outside of the rotor and is adjacent to the tidal flow is thermally beneficial.

3.6.2 Direct drive wave power

Because of the nature of the wave energy resource, direct drive generators often tend to be linear machines.[36] One of the best documented direct drive

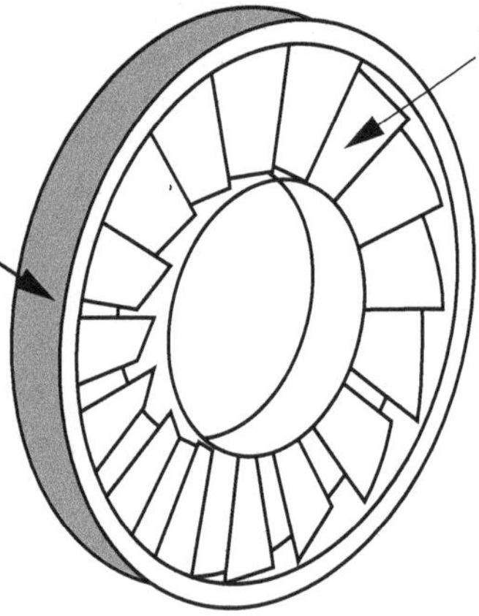

3.24 OpenHydro integrated tidal turbine/generator rotor with blades and magnets.[34,35]

wave energy converters is the Archimedes Wave Swing.[37] This 2 MW device used a permanent magnet linear generator mounted below a floater. As with other linear generators, special provisions must be introduced to limit the stroke of the device (so as to stop the translator and stator disengaging). The structural design[38] and the use of linear bearings to maintain the airgap between translator and stator[39] also pose challenges different to those in rotating electrical machines.

3.7 References

1. Polinder, H., de Haan, S.W.H., Dubois, M.R. and Slootweg, J.G.H., 'Basic operation principles and electrical conversion systems of wind turbines', *EPE Journal*, (2005), vol. **15**, no. 4, pp. 43–50.
2. Grauers, A., 'Design of direct-driven permanent-magnet generators for wind turbines', Technical report no 292, School of Electrical and Computer Engineering, Chalmers University of Technology, Göteborg, Sweden, (1996).
3. McDonald, A.S., Mueller, M.A. and Polinder, H., 'Comparison of generator topologies for direct-drive wind turbines including structural mass', Proc. Int. Conf. Electrical Machines, Chania, Crete, Greece, September (2006).
4. Hartkopf, T., Hofmann, M. and Jöckel, S., 'Direct-drive generators for megawatt wind turbines', *Proc. European Wind Energy Conf.*, Dublin, Ireland, October (1997), pp. 668–71.
5. Polinder, H., Bang, D.-J., van Rooij, R.P.J.O.M., McDonald, A.S. and Mueller, M.A., '10 MW wind turbine direct-drive generator with pitch or active speed stall', IEEE Int. Electric Machines and Drives Conf., Antalya, Turkey, May (2007).
6. Polinder, H., vander Pijl, F.F.A., de Vilder, G.-J. and Tavner, P.J., 'Comparison of direct-drive and geared generator concepts for wind turbines', *IEEE*

Transactions on Energy Conversion, vol. **21**, no. 3, pp. 725–33, September (2006). doi: 10.1109/TEC.2006.875476 URL: http://ieeexplore.ieee.org/stamp/stamp.jsp?tp=&arnumber=1677663&isnumber=35284.

7. Deok-je Bang; Polinder, H., Ferreira, J.A. and Seung-soo Hong, 'Structural mass minimization of large direct-drive wind generators using a buoyant rotor structure', *IEEE Energy Conversion Congress and Exposition (ECCE), 2010*, pp. 3561–8, 12–16 September (2010). doi: 10.1109/ECCE.2010.5617705 URL: http://ieeexplore.ieee.org/stamp/stamp.jsp?tp=&arnumber=5617705&isnumber=5617696.
8. GE 4.1-113 wind turbine, http://www.ge-energy.com/products_and_services/products/wind_turbines/fourone_113.jsp, last accessed July 2012.
9. MTorres wind, http://www.mtorres.es/pdf/TWT%20165-82%20IN.pdf, last accessed July 2012.
10. STX Windpower B.V., http://www.stxwind.com/nl/products/46-stx93_2_0, last accessed July 2012.
11. Bywaters, G., John, V., Lynch, J., Mattila, P., Norton, G., Stowell, J., Salata, M., Labath, O., Chertok, A. and Hablanian, D., 'Northern Power Systems WindPACT drive train alternative design study report', National Renewable Energy Laboratory, (2004).
12. Spooner, E., Gordon, P., Bumby, J.R. and French, C.D., 'Lightweight ironless-stator PM generators for direct-drive wind turbines', *IEE Proceedings of Electric Power Applications*, vol. **152**, no. 1, pp. 17–26, 7 January (2005). doi: 10.1049/ip-epa:20041084 URL: http://ieeexplore.ieee.org/stamp/stamp.jsp?tp=&arnumber=1395487&isnumber=30360.
13. Klontz, K.W., Miller, T.J.E., McGilp, M.I., Karmaker, H. and Zhong, P., 'Short-circuit analysis of permanent-magnet generators', *IEEE Transactions on Industry Applications*, vol. **47**, no. 4, pp. 1670–80, July–August (2011). doi: 10.1109/TIA.2011.2154370 URL: http://ieeexplore.ieee.org/stamp/stamp.jsp?tp=&arnumber=5766739&isnumber=5956885.
14. Tavner, P.J. and Spooner, E., 'Light structures for large low speed machines for direct drive applications', Proc. International Conference on Electrical Machines, Chania, Greece, (2006).
15. McDonald, A.S., 'Structural analysis of l Low speed, high torque electrical generators for direct drive renewable energy converters', Ph.D. dissertation, University of Edinburgh, Edinburgh, Scotland, (2008).
16. Mueller, M.A., McDonald, A.S. and Macpherson, D.E., 'Structural analysis of low-speed axial-flux permanent-magnet machines', *IEE Proceedings of Electric Power Applications*, (2005), vol. **152**, no. 6, pp. 1417–26.
17. ANSYS, http://www.ansys.com, last accessed July 2012.
18. McDonald, A.S., Mueller, M.A. and Polinder, H., 'Structural mass in direct-drive permanent magnet electrical generators', *IET Renewable Power Generation*, vol. **2**, no. 1, pp. 3–15, March (2008). doi: 10.1049/iet-rpg:20070071. URL: http://ieeexplore.ieee.org/stamp/stamp.jsp?tp=&arnumber=4463796&isnumber=4463794.
19. Chipperfield, A.J. and Fleming, P.J., 'The MATLAB® genetic algorithm toolbox', *IEEE Colloquium on Applied Control Techniques Using MATLAB*, Digest No. 1995/014, pp. 10/1–10/4, London, UK, 26 January (1995).
20. McDonald, A.S. and Mueller, M.A., 'Development of analytical tools for estimating inactive mass', University of Edinburgh, Scotland, UPWIND report, (2008).

21. Zavvos, A., McDonald, A.S. and Mueller, M., 'Electromagnetic and mechanical optimisation of direct-drive generators for large wind turbines', IET Power Electronics Machines and Drives Conference, Brighton, UK, (2010).
22. Zavvos, A., Bang, D.J., McDonald, A.S., Polinder, H. and Mueller, M., 'Structural analysis & optimisation of transverse flux permanent magnet machines for 5 MW and 10 MW direct drive wind turbines', *Journal of Wind Energy: The UpWind Special Issue*, vol. **15**, no. 1, pp. 19–43, January (2012).
23. Bang, D., 'Design of transverse flux permanent magnet machines for large direct-drive wind rurbines', Ph.D. dissertation, Delft University of Technology, Delft, The Netherlands, (2010).
24. Engstrom, S. and Lindgren, S., 'Design of NewGen direct drive generator for demonstration in a 3.5 MW Wind Turbine', European Wind Energy Conference, Milan, Italy, (2007).
25. Shrestha, G., Polinder, H., Bang, D.J., Ferreira, J.A. and McDonald, A.S., 'A new concept for weight reduction of large direct drive machines', *18th International Conference on Electrical Machines*, pp. 1–6, 6–9 September (2008). doi: 10.1109/ICELMACH.2008.4800206. URL: http://ieeexplore.ieee.org/stamp/stamp.jsp?tp=&arnumber=4800206&isnumber=4799813.
26. Pyrhönen, J., Jokinen, T. and Hrabovcova, V., '*Design of Rotating Electrical Machines*', Wiley-IEEE Press, (2008).
27. Vensys 1.5MW wind turbine, http://www.vensys.de/energy-en/produkte-und-service/vensys-1-5-mw.php, last accessed July 2012.
28. Vensys 2.5MW wind turbine, http://www.vensys.de/energy-en/produkte-und-service/vensys-2-5-mw.php, last accessed July 2012.
29. ENERCON GmbH, http://www.enercon.de/en/_home.htm, last accessed July 2012.
30. Semken, R.S., Polikarpova, M., Röyttä, P., Alexandrova, J., Pyrhönen, J., Nerg, J., Mikkola, A. and Backman, J., 'Direct-drive permanent magnet generators for high-power wind turbines: benefits and limiting factors', *IET Renewable Power Generation*, vol. **6**, no. 1, pp. 1–8, January (2012). doi: 10.1049/iet-rpg.2010.0191. URL: http://ieeexplore.ieee.org/stamp/stamp.jsp?tp=&arnumber=6141499&isnumber=6141498.
31. IEC 60085 standard, 'Electrical insulation – thermal evaluation and designation', 2007.
32. Mellor, P.H., Roberts, D. and Turner, D.R, 'Lumped parameter thermal model for electrical machines of TEFC design', *IEE Proceedings B – Electric Power Applications*, vol. **138**, no. 5, pp. 205–218, September (1991). URL: http://ieeexplore.ieee.org/stamp/stamp.jsp?tp=&arnumber=92929&isnumber=3045.
33. Motor Design, http://www.motor-design.com, last accessed July 2012.
34. Open Hydro, http://www.openhydro.com/home.html, last accessed July 2012.
35. Spooner, E., 'A Hydroelectric Turbine', patent WO 2008/006614 (A1), 17 January (2008).
36. Mueller, M.A., 'Electrical generators for direct drive wave energy converters', *IEE Proceedings – Generation, Transmission and Distribution*, vol. **149**, no. 4, pp. 446–56, July (2002). doi: 10.1049/ip-gtd:20020394. URL: http://ieeexplore.ieee.org/stamp/stamp.jsp?tp=&arnumber=1024191&isnumber=22022.
37. Polinder, H., Damen, M.E.C. and Gardner, F., 'Design, modelling and test results of the AWS PM linear generator', *European Transactions on Electrical Power*, vol. **15**, pp. 245–56, (2005). doi: 10.1002/etep.56.

38. McDonald, A.S., Crozier, R., Caraher, S., Mueller, M.A. and Chick, J.P., 'Integrated design of direct-drive linear generators for wave energy converters', *International Conference on Sustainable Power Generation and Supply*, pp. 1–7, 6–7 April (2009). doi: 10.1109/SUPERGEN.2009.5348025. URL: http://ieeexplore.ieee.org/stamp/stamp.jsp?tp=&arnumber=5348025&isnumber=5347868.
39. Caraher, S., 'Bearing options, including design and testing, for direct drive linear generators in wave energy converters', Ph.D. dissertation, University of Edinburgh, November (2011), http://www.era.lib.ed.ac.uk/handle/1842/5740.

4

An overview of power electronic converter technology for renewable energy systems

Z. CHEN, Aalborg University, Denmark

DOI: 10.1533/9780857097491.1.80

Abstract: This chapter presents power electronic technology which is an enabling tool for modern wind and marine energy conversion systems. In this chapter, the main power electronic devices are described. Various power electronic converter topologies are presented, and commonly used modulation schemes and control methods are introduced.

Key words: power electronic components, power electronic converters, modulation, control.

4.1 Introduction

Power electronic technology is used to convert electrical power from one form into another so as to efficiently match the application characteristics. The power electronic converter is an enabling technology for renewable energy power generation systems (Chen *et al.*, 2009). Most renewable energy-based power generation technologies, such as wind turbines and solar photovoltaics (PV) systems, normally have a set of optimal operational conditions (frequency, voltage, etc.) which presents good energy capture/conversion efficiency, but may not directly match the grid or consumer specifications. The power electronic converter is then used to provide the connection/conversion between the generation units and the grid/consumers, to achieve high efficiency and meet the grid requirements, including frequency, voltage, active and reactive power, flickers, harmonics and ride-through capabilities, etc. Power electronic converters are playing an increasingly significant role in the development of modern wind turbines and wind farms.

Power electronic technology, including semiconductor devices, circuit topologies, modulation and control methods, has seen rapid development in recent years. Performance is continuously being improved while costs are falling, and the application areas have increased significantly. This chapter introduces the basics of power electronic technologies, including semiconductor components, converter configurations, modulation and control methods.

This chapter is organized into the following sections: Section 4.2 introduces the main power electronic components; Section 4.3 presents the main topologies of power electronic converters; Section 4.4 discusses the modulation technology of voltage source converters (VSCs); and Section 4.5 describes power control of VSCs. Finally, a summary and discussion of future trends are given in Section 4.6.

4.2 Power electronic components

Power electronics has experienced revolutionary changes during the last three to four decades, and one of the important development areas is in semiconductor devices.

Semiconductors, being basic power electronic components and acting as switches in power electronic circuits, may be classified as non-controllable, AC voltage-commutated and self-commutated (switchable) semiconductors as shown in Fig. 4.1. Diodes are non-controllable devices in which on and off states are completely determined by external circuit conditions. Fast recovery diodes are used in some pulse width modulation (PWM) converters. The AC voltage-commutated component, the thyristor, relies on an external AC voltage to stop current conduction but the device needs a gate control to turn on. However, a self-commutated component may start/stop the current conduction by applying a voltage or current signal on the component gate.

The main important parameters of a semiconductor are the breakdown voltage and the rated current. Power electronic devices are still undergoing important developmental improvements. The breakdown voltage and/or the rated current of semiconductors are constantly increasing. Important research is also being conducted on the use of new materials, such as silicon carbide (SiC) (Hui Zhang and Tolbert, 2011), which could dramatically

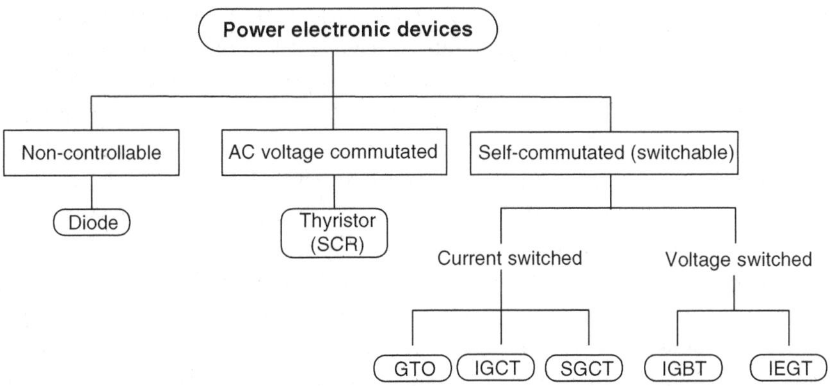

4.1 Classification of some medium- and high-power semiconductors.

Table 4.1 Characteristics of the main switchable semiconductor devices

Power semiconductor	Company	Ratings (V_{DRM}, I_{TGQM}/V_{CE}, I_c)	Switch power S_{max} (MVA)	Case
GTO	MITSUBISHI	6000 V; 6000 A asym.	36	Press-pack
		4500 V: 1000–4000 A asym.	18	Press-pack
	ABB	4500 V;	18	Press-pack
		600–4000 A asym.	18	Press-pack
		6000 V; 3000 A asym.		
IGBT	EUPEC	3300 V; 400–1200 A	3.96	Module
		6500 V; 200–600 A	3.90	Module
	MITSUBISHI	3300 V; 400–1200 A	3.96	Module
		4500 V; 400–900 A	4.05	Module
	HITACHI	3300 V; 400–1200 A	3.96	Module
	TOSHIBA	3300 V; 400–1200 A	3.96	Press-pack
		4500 V; 1200–2100 A	9.45	Module
	ABB	3300 V; 1200 A	3.96	Module
		4500 V; 600–3000 A	13.50	Press-pack
		6500 V; 600 A	3.90	Module
IGCT	ABB	4500 V; 3800–4000 A	18.00	Press-pack
		asym.	9.90	Press-pack
		4500 V; 340–2200 A r.c.	9.90	Press-pack
		5500 V; 280–1800 A r.c.	18.00	Press-pack
		6000 V; 3000 A asym.		
	MITSUBISHI	4500 V; 4000 A asym.	18	Press-pack
		6000 V; 3500–6000 A	36	Press-pack
		asym.	9.75	Press-pack
		6500 V; 400–1500 A sym.		

increase the power density of the converters. Recently, SiC devices have been developed with high power capacity and small size.

The self-commutated or switchable power semiconductor devices may be further divided into current-switched and voltage-switched devices, depending on the required gate drive characteristics. Turning a current-switched device on or off requires a gate current while turning a voltage-switched device on or off depends on a gate voltage. At the power level under consideration, current-switched devices mainly include IGCTs (Insulated Gate Commutated Thyristors), SGCTs (Symmetrical Gate Commutated Thyristors) and GTOs (Gate Turn-Off Thyristors), while the main voltage- switched devices are IGBTs (Insulated Gate Bipolar Transistors) and IEGTs (Injected Enhanced Gate Transistors). Current-switched devices may have lower conduction losses than voltage-switched devices, but have higher switching losses than voltage-switched devices. Higher switching speeds are possible with voltage-switched devices, but low power devices may have higher switching speed, such as MOSFET, and SiC FETs, etc. Table 4.1 presents the key rating parameters of some semiconductors.

Two of the most common switching devices are IGBTs and IGCTs. The IGCT has a higher maximum current rating, but its maximum switching frequency is lower. The higher switching speed of the IGBT can improve the harmonic performance at the expense of higher switching losses. The conduction (on state) voltage drop across IGBTs is higher than that with IGCTs.

IGBTs are widely used for power electronic converters and available at voltage ratings up to 4500 V current ratings up to 3000 A. Higher-voltage IGBTs are available, but normally with a reduced current capability (see Table 4.1). The IGCT is a device with a higher rating: state-of-the-art asymmetric blocking IGCT devices are available at 6 kV, 6000 A. Though IGCTs are suitable for higher-power applications, IGBTs are very attractive for medium-voltage applications due to the simplicity of the gate drive, protection (both over-current and over-voltage), possible snubberless operation and high switching speed.

Compared to GTOs, both IGBTs and IGCTs present better costs and power density performance; therefore, it is expected that GTOs may be replaced by IGBTs in most lower-medium-power applications, and by IGCTs in higher-power range.

Commercial semiconductors may be packaged into plastic-pack (module type) or press-pack. Plastic-pack IGBTs have single-sided cooling and behave as an open circuit in a failure with the possibility of case rupture and arc flash event. Press-pack IGBTs behave as a short circuit in the event of a failure. IGCTs may only be fabricated in press-pack technology. With press-pack technology, the cooling system can be designed to be double-sided so that the effectiveness of the external cooling system is almost doubled.

IGBT devices may also be made in Intelligent Power Module (IPM) form. An IPM is a hybrid device incorporating gate drivers with built-in protection features. It could result in a simple converter design.

IEGTs are generally manufactured with press-pack technology and have a lower forward voltage drop than IGBTs, which may lead to a higher power density and higher efficiency. IEGTs have a maximum voltage rating of 4.5 kV, an average current rating of 2.1 kA and a peak current rating of 4.2 kA. However, IEGTs require more components in their gate circuits and do not have the inherent over-current protection capability of IGBT devices (Badrzadeh *et al.*, 2009).

SGCT is a thyristor with a similar integrated gate drive to the IGCT. SGCTs can block voltage in both directions while allowing only one direction of current; therefore, they do not need a series diode when used as a switch in a current source converter (CSC), which reduces the number of components required. SGCTs are available with ratings of 6000 V and 5000 A, leading to device rating of 30 MVA. So far, SGCTs have only been employed for CSC (Badrzadeh *et al.*, 2009).

4.3 Topologies of power electronic converters

Since power electronic converters are constructed with semiconductor devices, driving, protection and control circuits, they perform the conversion and control of voltage/current magnitude and frequency. Depending on the particular devices used, the converter systems may be classified as shown in Fig. 4.2. The most common topology for three-phase application is the six-pulse bridge circuits shown in Fig. 4.3.

A diode rectifier can only convert AC power into DC power. The grid commutated converters with high power capacity thyristors are mainly used for very high voltage and power applications, such as conventional High Voltage Direct Current (HVDC) systems and large synchronous motor drives. A thyristor can transfer active power in both directions (AC→DC or DC→AC), but consumes inductive reactive power and is not able to control the reactive power.

Self-commutated converters use switchable semiconductors and normally adopt PWM switching methods. This type of converter may transfer active power in both directions (AC→DC or DC→AC), and can also control the AC side reactive power in both directions (inductive and capacitive).

The most common self-commutated converters are DC link converters, which may be further divided into two types, VSC and CSC, depending on the type of DC link. Typical circuits of VSC and CSC are illustrated in Fig. 4.3. A voltage source converter has a DC capacitor to ensure a smooth DC voltage. The AC output of a VSC is a PWM voltage waveform. A CSC has a DC reactor to maintain smooth direct current and generates a PWM current waveform at the AC terminal. Self-commutated VSC and CSC may perform four-quadrant operation (transferring active power in both directions and absorbing/producing reactive power at the AC side). The PWM

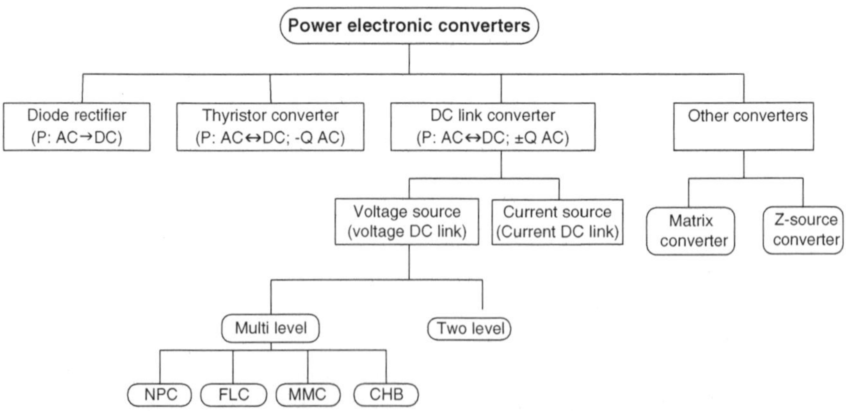

4.2 Classification of AC↔DC power electronic converters.

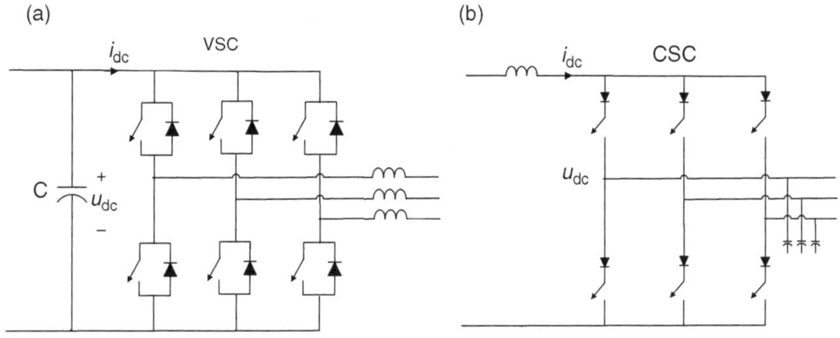

4.3 Schematic of six-pulse power electronic converters. (a) Three phase 2-level VSC. (b) Three phase 2-level CSC.

high- frequency switching techniques may produce harmonics and interharmonics in the range of some kHz. These harmonics are relatively easy to remove with small filters. At present, most power electronic converters used in wind power are VSCs.

Power electronic converters may be connected into various configurations, while two-level converters as shown in Fig. 4.3 are the simplest configuration.

The power electronic converters may be further configured for high power and lower harmonics operation, for example, multi-level configurations.

There are various multi-level converter circuit topologies, for example, the Neutral- Point Clamped (NPC) converter, Flying Capacitor (FLC) converter, Modular Multi-level Converters (M^2LC or M^2C or MMC) and Cascade H-Bridge (CHB) converter. The one-leg circuits of NPC, FLC and MMC converters are shown in Fig. 4.4a, 4.4b and 4.4c, and a single-phase circuit of a multi-level cascaded H-bridge inverter is shown in Fig. 4.4d. Multi-level converters can vary their output between several voltage levels, which can provide improved voltage waveforms, reduced filtering requirements and reduced common mode voltage and EMI (electromagnetic interference) impacts. In a multi-level converter, dynamic voltage sharing and balancing may be built into the structure through either clamping diodes or clamping capacitors, and be dealt with in coordination with control techniques.

The three-level NPC converter is now available with mainly IGBT or IGCT implementations, and may readily be applied for large wind turbine generators.

The FLC converter has been researched by Krug *et al.* (2003). The advantages offered by the FLC include equal load sharing between the switching devices and inherent redundancies in the switching states. The large DC link capacitance may contribute to the ride-through capabilities.

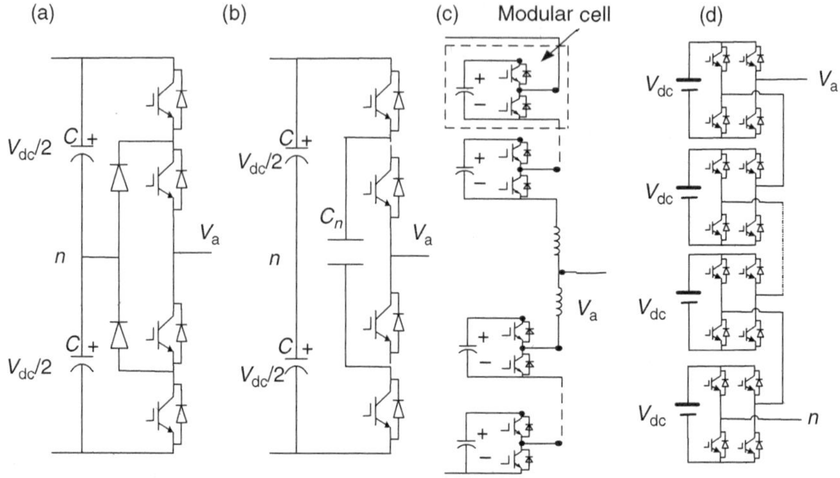

4.4 One-leg/single-phase multi-level VSCs. (a) One-leg of 3 level NPC VSC. (b) One-leg of 3 level FLC VSC. (c) One-leg of a modular multilevel converter (MMC) VSC. (d) Single-phase of a multilevel cascaded H-bridges inverter.

An MMC leg is constructed by connecting a number of converter cells (as shown in Fig. 4.4c) and two inductors. Each cell has its own capacitor which can be charged or discharged by controlling the relevant semiconductors of the cell. The capacitors in the cells could be considered as the distributed DC capacitors. MMC is proposed for high-power applications (Glinka and Marquardt 2003; Marquardt and Lesnicar, 2003). The MMC benefits from easy construction and assembly and flexibility in converter design, and is suitable for high- or medium-voltage power conversion. The MMC applications in STATCOM and VSC-HVDC have been described in Siemens' SVC PLUS and HVDC PLUS.

The cascaded H-bridge (CHB) converter, which is connected using a number of series-connected power cells, can achieve the increased voltage rating by increasing the number of series-connected cells. This arrangement requires a large number of switching devices. The converter footprint is large and the large number of devices may lead to reliability concerns. On the other hand, the CHB presents the lowest harmonic distortion and the lowest common mode voltage of all medium-voltage converters, so that an output filter may not be needed. The cascaded H-bridge configuration needs a number of DC voltage inputs and may not be suitable for a common DC link inverter configuration (Tolbert and Peng, 2000).

Examples of three-phase multi-level converters, diode-clamped NPC-VSC and start-configured cascaded H-bridges, are shown in Fig. 4.5.

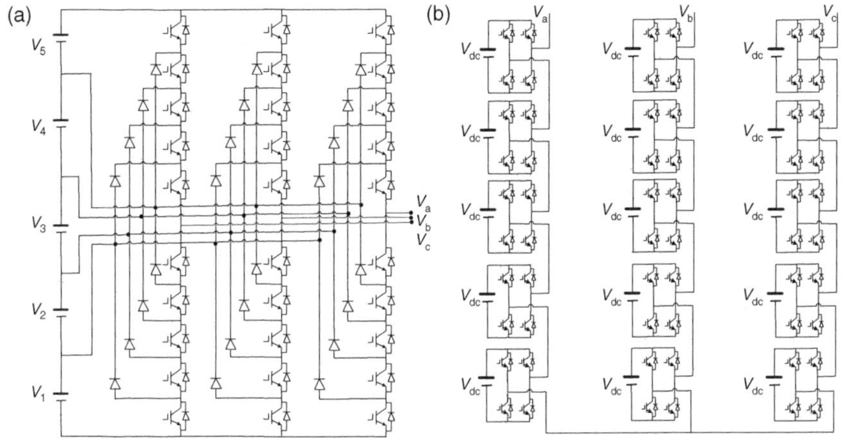

4.5 Three-phase multi-level voltage source converter examples. (a) Three phase 6 level diode-clamped NPC-VSC. (b) Three phase 11-level start-configured cascaded H-bridges.

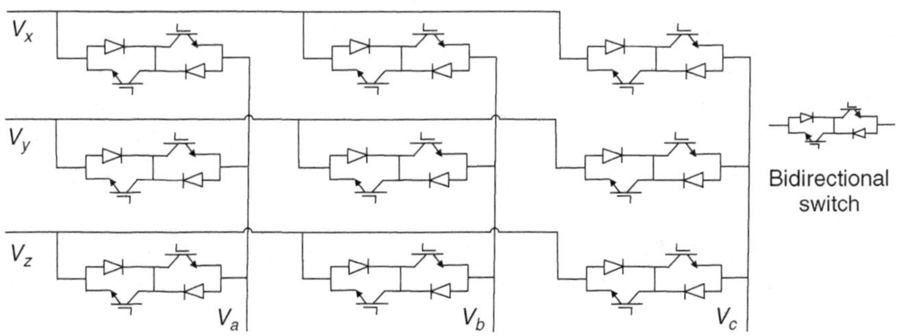

4.6 Circuit configuration of a matrix converter.

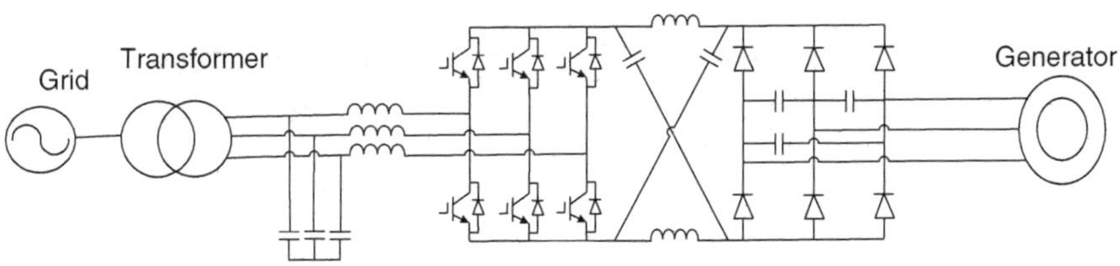

4.7 Circuit configuration of a Z-source converter. (Source: Reproduced based on Supatti and Peng, 2009.)

Other converter configurations are also proposed, such as matrix converters as shown in Fig. 4.6, and Z-source converters (Supatti and Peng, 2009) as shown in Fig. 4.7, respectively.

A matrix converter is an 'all silicon' AC/AC converter formed by bidirectional switches, without significant reactive elements. This converter may be a promising technology that could contribute to the development of multi-megawatt wind turbines (Helle *et al.*, 2004; Barakati *et al.*, 2009). Some possible technological barriers may be: the absence of tailor-made bidirectional switches, and poor ride-through capability against grid disturbances. Further research is needed.

The Power Electronic Building Block (PEBB), power devices integrated with gate drivers and other components into a building block, is a universal power processor that can transform electrical power input to an output with the desired form of voltage, current and frequency. A PEBB may reduce cost, losses, weight, size and engineering effort as well as maintenance cost and commissioning time; consequently, the complexities in the development and design of power electronics systems can be reduced (Ericsen *et al.*, 2006).

4.4 Modulation techniques in voltage source converters (VSCs)

Modulation is the process of switching the semiconductors in a power electronic converter from one state to another to produce a desired output.

PWM techniques play an important role in modern power electronics and affect the power losses, dynamic requirements, harmonics and EMI. Many PWM techniques have been developed. Some PWM methods for DC link VSCs are briefly introduced here.

The PWM methods for VSC applications may be classified as voltage-control PWM and current-control PWM schemes, as shown in Fig. 4.8. The current-control PWM scheme usually switches the semiconductors to generate a desired AC current waveform. On the other hand, the voltage control methods produce a controllable AC voltage fundamental component.

4.4.1 Voltage-control pulse width modulation (PWM) techniques for VSCs

As Fig. 4.8 shows, the voltage-control PWM modulation methods may be classified as random PWM and fixed switching frequency PWM, which includes conventional PWM techniques, programmed angle PWM techniques and Space Vector PWM (SVPWM), etc.

The simplest scheme is the natural sinusoidal PWM, one of the conventional PWM techniques, where a sinusoidal reference signal (V_{ref}) is compared with a triangular waveform (V_{tri}), and the semiconductor switching is determined by the crossing points of the reference signal and the triangular waveform. The frequency of the triangular waveform establishes the

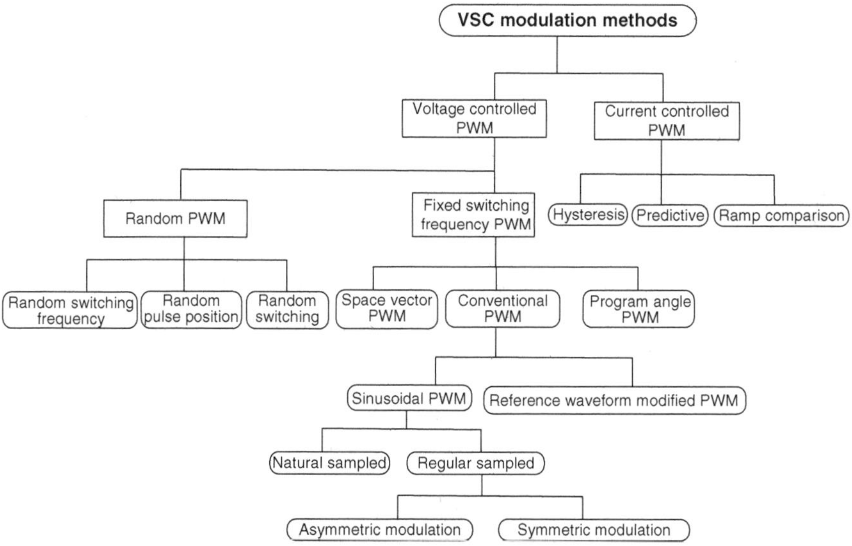

4.8 Modulation methods for voltage source converter (VSC).

converter switching frequency (f_{sw}). The frequency of the reference signal determines the converter AC side output frequency. The magnitude of the fundamental component can be controlled by regulating the modulation ratio (M_a) which is the ratio of peak values V_{ref} to V_{tri}.

$$M_a = \frac{\hat{V}_{ref}}{\hat{V}_{tri}} \quad [4.1]$$

Example waveforms of a three-phase sinusoidal PWM are shown in Fig. 4.9 (Chen, 1997) where f_{sw} = 750 Hz and M_a = 0.8 are chosen for a clear illustration.

Various switching techniques have been proposed based on the modification of the sinusoidal PWM scheme, such as conventional sinusoidal PWM with regular sampling (symmetric or asymmetric), the modified sine PWM, harmonic injection PWM, etc. For example, a harmonic injection PWM takes advantage of the absence of a triple current path in a system without a neutral connection to inject triple harmonics into the modulating waveform, thus increasing the fundamental frequency component of the converter output voltage.

With the development of numerical techniques and microprocessors, programmed angle PWM techniques have also been developed. In a programmed angle PWM switching technique, a group of preferred switching angles could be pre-calculated to optimize a certain performance index, such as selective harmonic elimination.

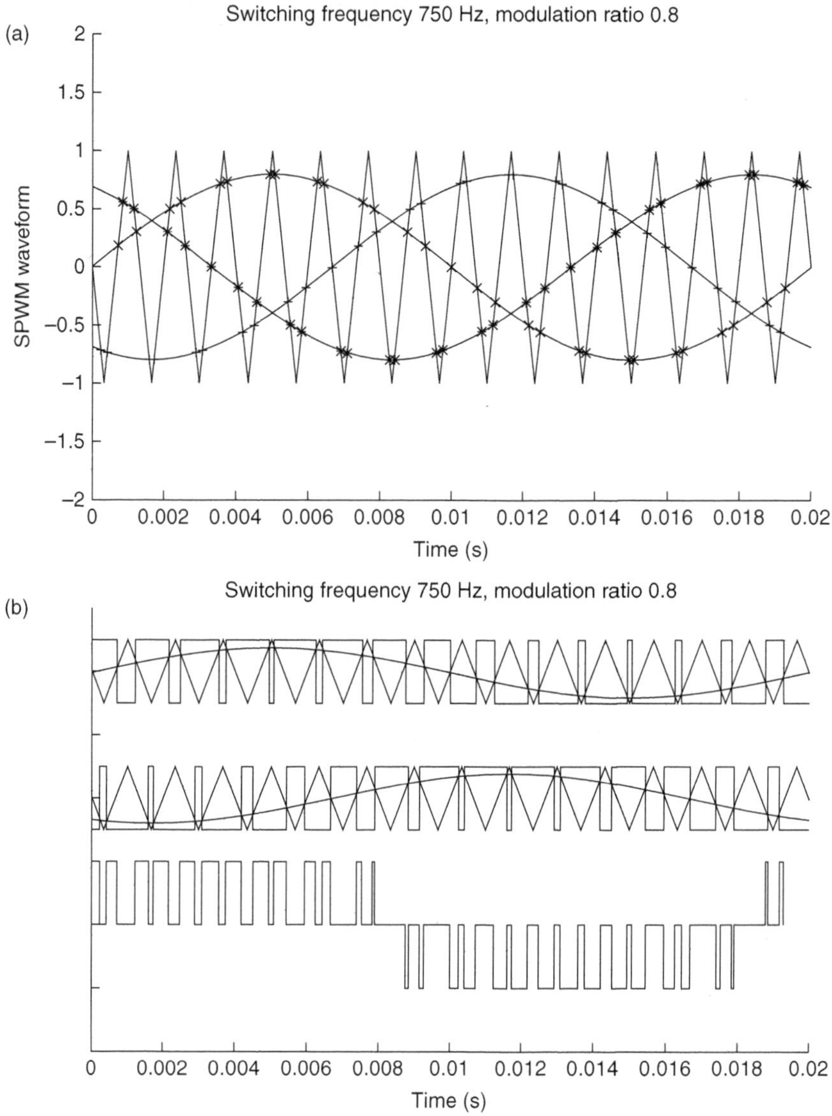

4.9 Natural sampled sinusoidal PWM waveforms (Chen, 1997). (a) Reference and triangular waveforms. (b) Switching signals and line to line voltage waveform.

The SVPWM approach demonstrates the advantage of higher efficiency over other methods. The principle of the SVPWM is illustrated in Fig. 4.10. The desired voltage, V_{des}, can be expressed in space-vector form as a combination of the converter three-phase output voltage v_a, v_b and v_c (harmonics are ignored):

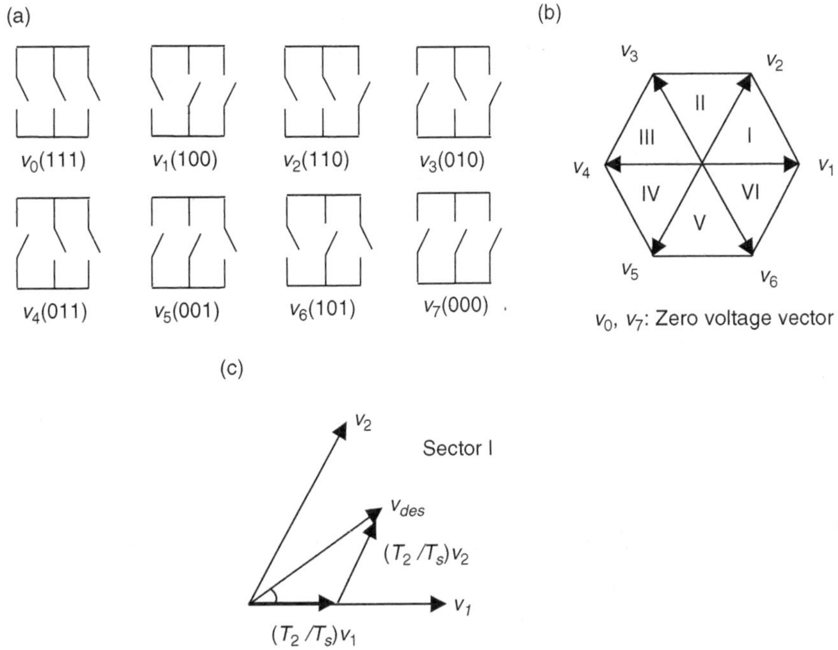

4.10 Space voltage vector PWM scheme. (a) Eight switch positions, (b) basic voltage vectors and (c) voltage vector composition.

$$\bar{V}_{\text{des}} = \sqrt{\frac{2}{3}}\left(v_a + av_b + a^2 v_c\right) \qquad [4.2]$$

where

$$a = e^{j120^\circ}$$
$$v_a = V_m \sin\omega t$$
$$v_b = V_m \sin(\omega t - 120^\circ)$$
$$v_c = V_m \sin(\omega t + 120^\circ)$$

and V_m is the amplitude of the fundamental component.

For a three-phase two-level VSC, there are eight possible switching positions, as shown in Fig. 4.10a. The corresponding voltage vectors are depicted in Fig. 4.10b.

A desired voltage vector may be synthesized by using two of the basic vectors as shown in Fig. 4.10c. The space vector PWM strategy for VSCs can be realized by selecting the appropriate switching vectors and determining their corresponding dwelling periods. If the reference vector is located in sector *I* as shown in Fig. 4.10c, then it may be composed of voltage vectors

v_1, v_2, and zero-voltage vectors v_0 and v_7. The relationships between the desired voltage vector and its two corresponding basic voltage vectors can be expressed as

$$\bar{V}_{\text{des}} = v_1 \frac{T_1}{T_s} + v_2 \frac{T_2}{T_s} \qquad [4.3]$$

where T_s is the period for $\bar{V}_{\text{des}}$ and T_1 and T_2 are the dwelling time for v_1 and v_2, respectively. This voltage space vector can be described in rectangular co-ordinates as:

$$T_1 \cdot \sqrt{\frac{2}{3}} V_d \cdot \begin{bmatrix} 1 \\ 0 \end{bmatrix} + T_2 \cdot \sqrt{\frac{2}{3}} V_d \cdot \begin{bmatrix} \cos 60^\circ \\ \sin 60^\circ \end{bmatrix} = T_s \cdot \sqrt{\frac{2}{3}} V_d \cdot k \cdot \begin{bmatrix} \cos \theta \\ \sin \theta \end{bmatrix} \qquad [4.4]$$

where $k = |V_{\text{des}}| / \sqrt{2/3} V_d$, $0 \le \theta \le 60^0$, and V_d is the DC link voltage. T_1, T_2 and T_0 can be solved from Equation [4.4] as

$$T_1 = T_s \cdot k \cdot \frac{\sin(60 - \theta)}{\sin 60^0}$$

$$T_2 = T_s \cdot k \cdot \frac{\sin \theta}{sin 60^0}$$

$$T_7 = T_0 = \frac{T_s}{2} - T_2 - T_1$$

The cycle time should preferably be as short as possible so that the maximum deviation between the reference voltage vector and average voltage formed by combining the voltage vectors is minimized. However a higher frequency results in high switching power loss. For a certain sampling frequency, the performance can be improved if the following conditions are met:

- The cycle consists of only three successive switching states: two voltage vectors adjacent to the reference vector and a zero-voltage vector.
- The switching sequence of each converter leg is arranged in such a way that the transition from one state to another is performed by switching only one converter switch, in order to minimize switching frequency and power loss.
- The switching of the converter starts from one zero state and ends at the other zero state.

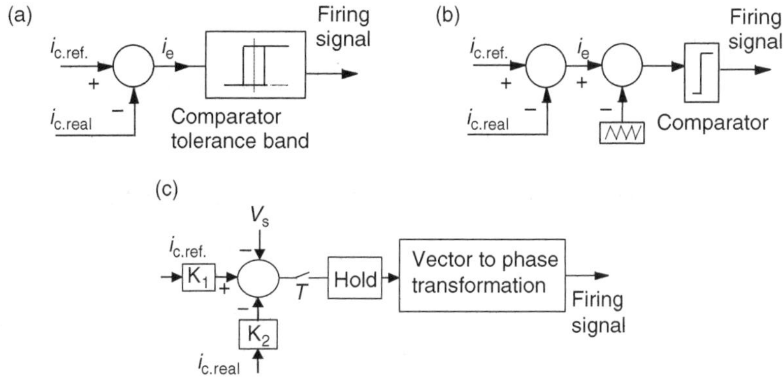

4.11 Controllers of current controlled VSC (Chen, 1997). (a) Hysteresis controller, (b) ramp comparison controller and (c) predictive controller.

4.4.2 Current-control PWM techniques for VSCs

A VSC may also be controlled to generate a desired current waveform at the AC side. In order to generate such a current waveform, the VSC should be operated in a closed-loop current-control mode which could offer good dynamic performance. Current controllers for the VSC may be divided into three basic types: hysteresis, ramp comparison and predictive controllers whose block diagrams are shown in Fig. 4.11 (Chen, 1997). The desired current, $i_{c,\text{ref}}$, and actual current, $i_{c,\text{real}}$, are compared and the error is used to drive the VSC.

Hysteresis controllers, shown in Fig. 4.11a, utilize some type of hysteresis in the comparison of the measured currents to the current references. When the line current becomes greater (less) than the current reference by the hysteresis band, the converter leg is switched to the negative (positive) DC direction to drive the current back into the band. The hysteresis band specifies the maximum current ripple, and the converter switching frequency varies over a cycle.

The ramp comparison controller compares the current error to a triangle waveform to generate the converter firing signals as shown in Fig. 4.11b. The switching signal is generated in a similar way to SPWM but the current error is used to replace the sinusoidal reference waveform. With this type of control, the converter is switched at the frequency of the triangle wave and its output current contains well-defined harmonics.

The block diagram of a predictive controller is shown in Fig. 4.11c. This controller calculates the converter voltage vector required to force the current to follow the current reference for every sample period. The switching frequency of the controller is well defined, but more system information and complex calculations are required. The calculation may delay the system response.

Compared with the complexity of the predictive controller and the switching frequency variation of the hysteresis controller, the ramp comparison controller is simple and gives a well-defined output harmonic distribution.

4.5 Power control of voltage source converters

The main purpose of controlling a power conversion system is to deliver the desired active and reactive power at the required specifications (voltage, frequency and other quality requirements), which can be implemented based on the modulation techniques discussed in Section 4.4. The power control of a power electronic converter may act in different ways. A common approach is the vector-control method based on a rotational reference frame.

4.5.1 Reference frames for vector control

Vector-control techniques have been well developed for PWM converters. Various reference frames may be used for vector-control applications. Figure 4.12 shows three sets of reference frames: *a-b-c* three-phase stationary reference frame, α-β two-axis stationary reference frame, and *d-q* two-axis rotational reference frame rotating at speed ω.

A set of balanced three-phase variables (f_a, f_b, f_c) in *a-b-c* reference frame, which could be currents or voltages or fluxes, may be respectively represented by the variables (f_α, f_β) in a α-β two-axis stationary reference frame with Equation [4.5] or the variables (f_d, f_q) in a *d-q* two-axis rotational reference frame with Equations [4.6] and [4.7].

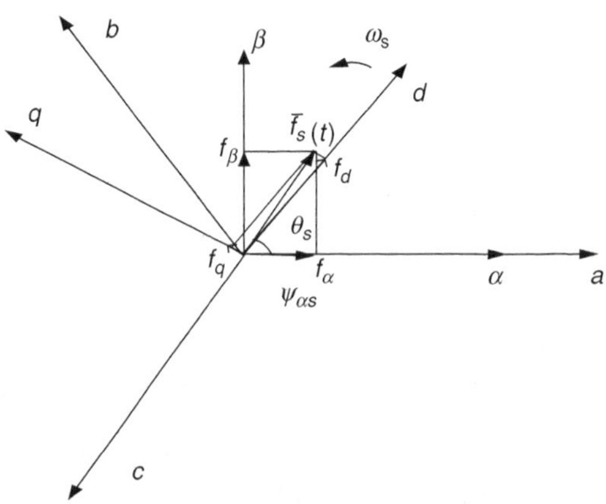

4.12 Example of reference frames.

$$\begin{bmatrix} f_\alpha \\ f_\beta \end{bmatrix} = \frac{2}{3}\begin{bmatrix} 1 & -1/2 & -1/2 \\ 2 & -\sqrt{3}/2 & \sqrt{3}/2 \end{bmatrix}\begin{bmatrix} f_a \\ f_b \\ f_c \end{bmatrix} \quad [4.5]$$

$$\begin{bmatrix} f_d \\ f_q \end{bmatrix} = \frac{2}{3}\begin{bmatrix} \cos(\theta) & \cos(\theta-2\pi/3) & \cos(\theta+2\pi/3) \\ \sin(\theta) & \sin(\theta-2\pi/3) & \sin(\theta+2\pi/3) \end{bmatrix}\begin{bmatrix} f_a \\ f_b \\ f_c \end{bmatrix} \quad [4.6]$$

Where

$$\theta = \int \omega dt = \tan^{-1}\frac{f_\beta}{f_\alpha} \quad [4.7]$$

The vector-control techniques can decouple the control of the active power and reactive power with a rotating *d-q* reference frame. A voltage is 90° ahead of its associated flux. The steady state *d-q* current components of an alternating current with a frequency of ω, i_d and i_q, will become DC components in a *d-q* reference frame rotating at the same speed of ω. The current component in the flux axis represents reactive power, while the current component in the voltage axis leads to active power. If the AC flux is fixed to the *d*-axis of the rotating reference frame (which means the projection of AC flux on the *q*-axis will be zero), the AC voltage is then located on the *q*-axis direction, so that i_q is related to the active power and i_d is associated with the reactive power. Otherwise, if the voltage is aligned to the *d*-axis of the reference frame, i_d is related to the active power and i_q is associated with the reactive power.

Various variables may be used for orientation, such as AC voltage-oriented reference frame, virtual flux-oriented reference frame, machine rotor-oriented reference frame, etc.

4.5.2 AC voltage-oriented vector control

A PWM voltage source converter connected to an AC source is used for illustration as shown in Fig. 4.13, where u_{sabc} are the three-phase AC source voltages, which may be represented by a space-vector voltage, u_s, as shown in Fig. 4.14, u_{Cabc} are the three-phase converter voltages, i_{sa}, i_{sb}, i_{sc} are the three-phase currents (represented by a space-vector voltage, i_s,) from the converter to the AC source, R_g and L_g are the inductor resistance and inductance, *C* is the DC link capacitor, i_{dcC}, i_{dcR} are the DC currents as shown in Fig. 4.13.

The voltage equations for the circuit shown in Fig. 4.13 are

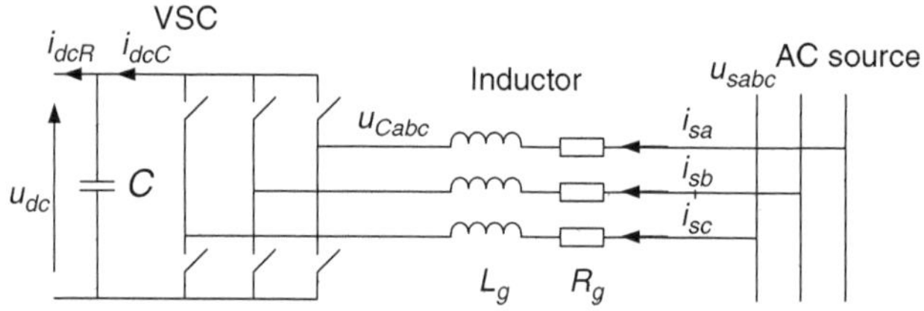

4.13 A voltage source converter connected to an AC source.

$$\begin{bmatrix} u_{sa} \\ u_{sb} \\ u_{sc} \end{bmatrix} = R_g \begin{bmatrix} i_{sa} \\ i_{sb} \\ i_{sc} \end{bmatrix} + L_g \frac{\mathrm{d}}{\mathrm{d}t} \begin{bmatrix} i_{sa} \\ i_{sb} \\ i_{sc} \end{bmatrix} + \begin{bmatrix} u_{Ca} \\ u_{Cb} \\ u_{Cc} \end{bmatrix} \qquad [4.8]$$

Using the *abc*-to-*dq* transformation [4.6], the corresponding equations in the *d-q* reference frame rotating at ω_s [rad/s] ($\omega_s = 2\pi f_s$, f_s is the electrical frequency of the AC source) are

$$\begin{aligned} u_{sd} &= R_g i_{sd} + L_g \frac{\mathrm{d}i_{sd}}{\mathrm{d}t} - \omega_s L_g i_{sq} + u_{Cd} \\ u_{sq} &= R_g i_{sq} + L_g \frac{\mathrm{d}i_{sq}}{\mathrm{d}t} + \omega_s L_g i_{sd} + u_{Cq} \end{aligned} \qquad [4.9]$$

where u_{sd}, u_{sq} are the grid voltages, u_{Cd}, u_{Cq} are the converter voltages, i_{sd}, i_{sq} are the currents, all in *d-q* axis frame. The active P_g and reactive Q_g power entering the AC source system are

$$\begin{aligned} P_g &= \frac{3}{2}\left(u_{sd} i_{sd} + u_{sq} i_{sq}\right) \\ Q_g &= \frac{3}{2}\left(u_{sq} i_{sd} - u_{sd} i_{sq}\right) \end{aligned} \qquad [4.10]$$

The angular position θ_s of the *d-q* reference frame is calculated as

$$\theta_s = \int \omega_s \mathrm{d}t = \tan^{-1} \frac{u_{s\beta}}{u_{s\alpha}} \qquad [4.11]$$

where $u_{s\alpha}$ and $u_{s\beta}$ are the ac source voltage components in the stationary $\alpha\beta$ axes.

Aligning the d-axis of the reference frame along the AC source voltage position given by Equation [4.11], the vector components of voltages and currents could be positioned as illustrated in Fig. 4.14, u_{sq} is zero, and u_{sd}

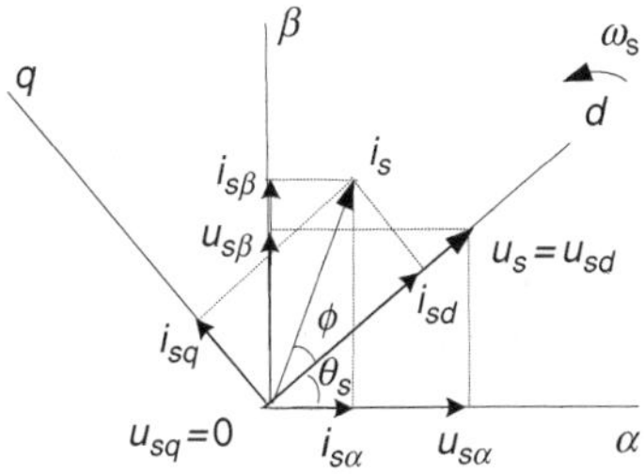

4.14 Vector diagram of VOC on *d-q* reference frame (AC voltage on *d*-axis).

becomes equal to the amplitude of the AC source voltage, u_s. Then the active and reactive power flow between the AC source and the converter will be proportional to i_{sd} and i_{sq} respectively, as given in Equation [4.12]. Therefore, the P_g and Q_g can be respectively controlled by controlling i_{sd} and i_{sq}.

$$\begin{aligned} P_g &= \frac{3}{2} u_{sd} i_{sd} \\ Q_g &= -\frac{3}{2} u_{sd} i_{sq} \end{aligned} \qquad [4.12]$$

The currents i_{sd} and i_{sq} can be controlled by converter voltages, u_{Cd} and u_{Cq} respectively. Considering Equation [4.9], the required voltage u^*_{Cd} and u^*_{Cq} can be expressed as follows:

$$\begin{aligned} u^*_{Cd} &= -u'_{Cd} + \left(u_{sd} + \omega_s L_g i_{sq}\right) \\ u^*_{Cq} &= -u'_{Cq} + \left(u_{sd} - \omega_s L_g i_{sd}\right) \end{aligned} \qquad [4.13]$$

$$\begin{aligned} u'_{Cd} &= R_g i_{sd} + L_g \frac{\mathrm{d}i_{sd}}{\mathrm{d}t} \\ u'_{Cq} &= R_g i_{sq} + L_g \frac{\mathrm{d}i_{sq}}{\mathrm{d}t} \end{aligned} \qquad [4.14]$$

Cascade control is normally used in a vector-control scheme. Usually the inner control loop is designed with a shorter time constant than the outer loop. Consequently, it is possible to design the outer control loop whilst neglecting the dynamics of the inner control loop. This assumption will ease the controller design. The proportional integral (PI)-controllers, usually used due to their simplicity and robustness, may be designed using standard controller design methods.

The vector-control scheme with the AC source voltage vector being aligned on the d-axis is also called voltage-oriented control (VOC). Block diagrams of VOC are shown in Fig. 4.15, where u^*_{Cabc} are the desired three-phase converter voltages, u^*_{Cd} and u^*_{Cq} are the desired converter voltages in d-q axis frame, i^*_{sd}, i^*_{sq} are the reference of the AC source currents in d-q axis frame. The i_{sq} determines the reactive power flow between the grid and the converter. If a zero reactive power exchange between the grid and the converter is preferred, the reference value, i^*_{sq}, may be set to zero. In Fig. 4.15b, a DC voltage control may also be added into the active current reference component calculation to help maintain a constant DC voltage.

4.5.3 Virtual flux-oriented reference frame

The voltages in Fig. 4.13 may be considered as being associated with fluxes, the virtual fluxes, which are the integral of the voltage and 90° behind the voltage. Equation [4.8] may be represented in space-vector format as

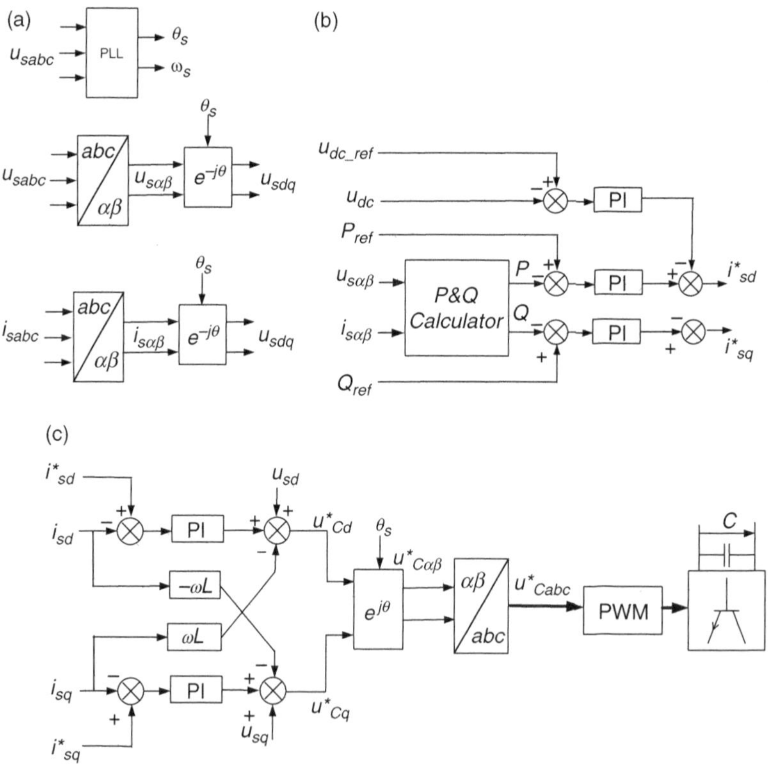

4.15 Block diagrams of VOC on the d q-reference frame. (a) PLL block, conversion of three phase voltages into dq voltages, and conversion of three phase currents into dq currents; (b) controllers for dq reference currents and (c) controllers for dq reference voltages and PWM generator.

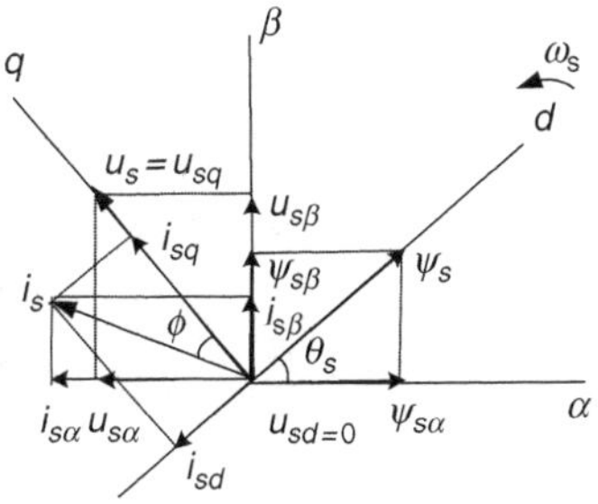

4.16 Vector diagram of flux-oriented control (FOC) on the *d*-*q* reference frame (virtual flux on *d*-axis).

$$u_s = R_g i_s + L_g \frac{di_s}{dt} + u_C \tag{4.15}$$

$$\psi_s = \int u_s dt \text{ and } \psi_c = \int u_s dt \tag{4.16}$$

Neglecting the series resistance of the line inductor, integrating [4.15] on both sides yields

$$\psi_s = L_g i_s + \psi_c \tag{4.17}$$

The flux angular position θ_s may be calculated from

$$\theta_s = \int \omega_s dt = \tan^{-1}\left(\frac{\psi_{s\beta}}{\psi_{s\alpha}}\right) \tag{4.18}$$

where $\psi_{s\alpha}$ and $\psi_{s\beta}$ are the flux components in $\alpha\beta$-axis reference frame ω_s is the electrical angular velocity of the AC source virtual flux (the frequency of voltage space vector, u_s).

With the virtual flux concept, aligning the *d*-axis of the reference frame along the AC source virtual flux position, as shown in Fig. 4.16, $\psi_{sd} = |\psi_s|$, $\psi_{sq} = 0$. The control system can also be designed on this reference frame, which is called the virtual flux-oriented control (VFOC). Figure 4.17 gives the general diagrams of VFOC, which have the same current controller structure as shown in Fig. 4.15c. The only differences are the calculations of reference angle, θ_s, and the i^*_{sd}, i^*_{sq}; these are shown in Fig. 4.17a and 4.17b as a comparison with those in Fig. 4.15a and 4.15b.

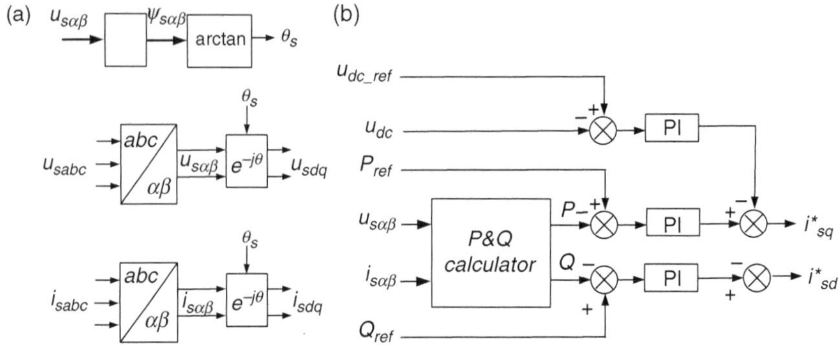

4.17 Block diagrams of voltage oriented control (VFOC) on the d q-reference frame. (a) Calculation of dq direction, conversion of three phase voltages into dq voltages, and conversion of three phase currents into dq currents; (b) controllers for dq reference currents.

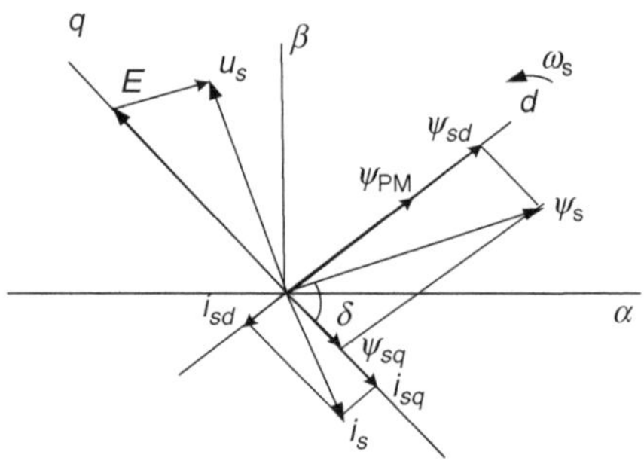

4.18 Diagram of rotor flux-oriented vectors.

4.5.4 Rotor flux-oriented vector control for synchronous machines

An electrical machine can also be controlled in a rotating dq-axis frame in a similar way to the discussion above. Aligning the *d*-axis of the reference frame to the rotor of a synchronous machine is to align the *d*-axis to the rotor flux rotating at the synchronous speed, ω_s. Taking a permanent magnet synchronous generator (PMSG) as an example, if the permanent magnet flux vector, Ψ_{PM}, is aligned on the *d*-axis, the electromotive force E is then on the *q*-axis, 90° ahead of the flux as shown in Fig. 4.18.

The rotational speed of the rotor flux, ω_s, is equal to the product of the rotor mechanical speed, ω_m, and the number of pole pairs, $p/2$:

$$\omega_s = \frac{p}{2}\omega_m \qquad [4.19]$$

The stator voltage equations in a *d-q* reference frame may be expressed as

$$
\begin{aligned}
u_{sd} &= R_s i_{sd} - \omega_s \psi_{sq} + \dot{\psi}_{sd} \\
u_{sq} &= R_s i_{sq} + \omega_s \psi_{sd} + \dot{\psi}_{sq}
\end{aligned} \qquad [4.20]
$$

The stator flux components are

$$
\begin{aligned}
\psi_{sd} &= L_d i_{sd} + \Psi_{\mathrm{PM}} \\
\psi_{sq} &= L_q i_{sq}
\end{aligned} \qquad [4.21]
$$

where u_{sd}, u_{sq} are the generator stator terminal voltage, ψ_{sd}, ψ_{sq} are the stator flux, i_{sd}, i_{sq} are the stator currents, L_d and L_q are the machine inductances, R_s is the stator winding resistance, all in the d-q reference frame, Ψ_{PM} is the magnitude of the flux produced by the permanent magnets of the rotor (the reference frame d-axis is aligned with the vector of the permanent magnet flux Ψ_{PM}), and ω_s is the electrical speed of the machine.

The electromagnetic torque can be expressed in the d-q reference frame as

$$
T_e = \frac{3}{2}\frac{p}{2}\left(\psi_{sd} i_{sq} - \psi_{sq} i_{sd}\right) \qquad [4.22]
$$

Substituting Equation [4.21] into Equation [4.22] gives

$$
T_e = \frac{3}{2}\frac{p}{2}\left(\Psi_{\mathrm{PM}} i_{sq} + \left(L_d - L_q\right) i_{sd} i_{sq}\right) \qquad [4.23]
$$

In Equation [4.23], the first term in the torque equation is produced by the interaction between the rotor magnetic field and i_{sq}, and the second term is the reluctance torque. If the permanent magnet generator is assumed to be a round rotor machine, which is a good approximation for the surface-mounted PMSG, then $L_d = L_q$, the electromagnetic torque of the generator may be expressed as

$$
T_e = \frac{3}{2}\frac{p}{2}\left(\Psi_{\mathrm{PM}} i_{sq}\right) \qquad [4.24]
$$

It is clear that i_{sq} is the torque current component, and i_{sd} is the magnetic field current component, which can be controlled respectively in a similar way to the VOC scheme. The control method is also called rotor flux-oriented control (RFOC).

VOC, VFOC and RFOC methods may all use an inner current-control loop with a structure shown in Fig. 4.15. Consequently, the control system structure and performance largely depend on the adopted current controllers. In addition to the PI controllers shown in Fig. 4.15, other types of current controllers may also be used, including hysteresis controllers and dead-beat controllers.

4.5.5 Direct torque control (DTC) and direct power control (DPC) methods

The control methods (VOC, VFOC and RFOC) mentioned above all use current vector controllers to control the decoupled d and q current components. The performance of such a converter largely depends on the accuracy of the current decoupling, the design of the current controllers and tuning of the parameters.

As alternatives to the current loop control method, some other control strategies have been proposed, for example, direct power control (DPC) which has been developed based on the direct torque control (DTC).

DTC is used to eliminate the need for a current regulator and provides direct control of a machine's torque and flux (Habetler *et al.*, 1992). A simple DTC has two reference signals, the stator flux amplitude and the electromagnetic torque, which are compared with the measured/estimated values respectively; the resultant errors are used as inputs to hysteresis controllers. The output of the hysteresis controllers are digital values which are used in a look-up table. The estimated flux position is also used for looking up the switching table. The driving signals for the power electronic converter are generated from the switching table.

Based on the DTC principle, DPC was developed for three-phase DC/AC converters (Noguchi *et al.*, 1998; Serpa *et al.*, 2005, Zhi *et al.*, 2009). Instead of torque and flux, DPC directly controls the active and reactive power without any internal control loop. The coordinate transformation and the decoupling between active and reactive power components are not needed. The switching states are selected from a switching table by using the instantaneous error between the measurement/estimation and the desired active and reactive power references. The power estimation can also be performed based on either voltage or virtual flux. Figure 4.19 shows a block diagram of a virtual flux-based DPC method.

The power can be expressed with the AC source flux and the converter flux as (Zhi *et al.*, 2009)

$$\begin{aligned} P_g &= -\frac{3}{2L_g}\omega_s \psi_{sd}\psi_{cq} \\ Q_g &= \frac{3}{2L_g}\omega_s \psi_{sd}\left(\psi_{cq} - \psi_{sd}\right) \end{aligned} \qquad [4.25]$$

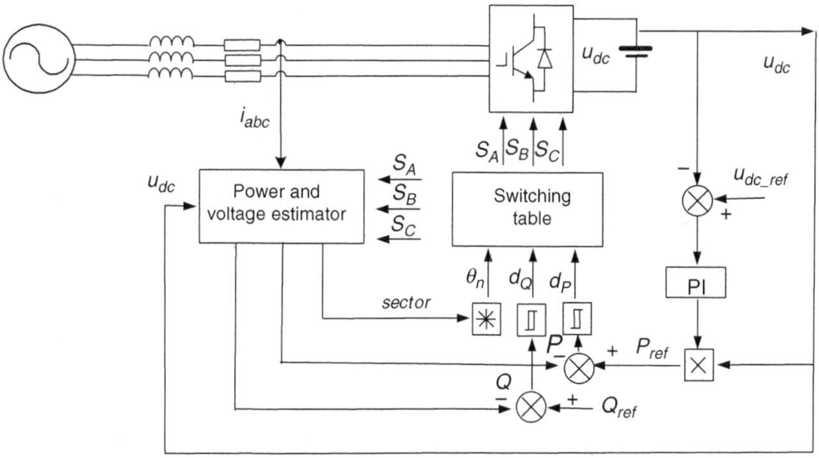

4.19 Block diagram of direct torque (power) control. (*Source*: Reproduced based on Serpa *et al.*, 2005.)

where ψ_{cd} and ψ_{cq} are the converter flux *d-q* components. Then the changes of power can be related to the changes of the converter fluxes as

$$\begin{aligned} \Delta P_g &= -\frac{3}{2L_g}\omega_s \psi_{sd}\Delta\psi_{cq} \\ \Delta Q_g &= \frac{3}{2L_g}\omega_s \psi_{sd}\Delta\psi_{cd} \end{aligned} \qquad [4.26]$$

The desired variation of the converter flux ($\Delta\psi_{cd}$ and $\Delta\psi_{cq}$) can be realized by the appropriate converter output voltage vector, u_c, as

$$\Delta\psi_c = u_c \Delta T$$

where ΔT is the time period. Consequently, a look-up table can be built to perform the direct power control. DPC approaches have some advantages, such as a fast dynamic response and simplicity. The disadvantages of the traditional DPC, such as the variable switching frequency, can be overcome by later proposed various versions of DPCs.

In this sub-section, the presented controllers are mainly based on rotating *d-q* reference frames, though the control may also be conducted on a stationary-axis frame or abc-axis frame with proportional resonant (PR) controllers (Blaabjerg *et al.*, 2006), which will not be described here due to space limitation.

4.6 Conclusion

This chapter has presented the basic power electronic technologies, including power electronic components, various power electronic converter topologies, and common modulation and control methods for voltage source converters (VSCs).

Research on power electronic components is continuing; new materials such as SiC-based devices have good potential to construct more efficient and compact power electronic converters with reduced filtering and cooling demands. Consequently, the new materials-based power semiconductors could have significant impacts on power electronic converter topology and modulation schemes.

VSCs are commonly used for interfacing renewable energy-based power generation systems. There are no barriers for up-scaling VSCs into next-generation large wind and marine power generation systems, though different topologies would provide different technical performances and incur different costs. An application-oriented optimization is needed to determine a cost-effective power electronic system for a particular system (see Chapter 5).

Neutral-point clamped converters are becoming standard topology in the industry, presenting good controllability and good performance under grid-faults, though redundancy may be required against semiconductor breakdown issues. A matrix converter has no mid-DC link passive components. This could bring advantages, such as avoiding the DC capacitor fail problem, however, the fault-ride-through capability may be a challenge. Multi-converter topologies, such as MMC and CHB, are suitable for higher-voltage and higher-power applications, and can provide good quality waveforms without high frequency switching, but the circuits and control are more complex.

4.7 References

Badrzadeh, B., Smith, K.S. and Wilson, R.C. (2009), 'Alternatives for high-power power electronic converters, switching devices and electric machines for very large wind farms: A technological and market assessment', EWEC2009 proceedings.

Barakati, S.M., Kazerani, M.A. and Aplevich, J.D. (2009), 'Maximum Power Tracking Control for a Wind Turbine System Including a Matrix Converter', *IEEE Transactions on Energy Conversion*, Vol. **24**, No. 3, pp. 705–13, September 2009.

Blaabjerg, F., Teodorescu, R., Liserre, M., and Timbus, A.V. (2006), 'Overview of control and grid synchronization for distributed power generation systems', *IEEE Transactions on Industrial Electronics*, Vol. **53**, No. 5, pp. 1398–409, October 2006.

Chen, Z. (1997), 'Advanced Wind Energy Converters Using Electronic Power Conversion', Ph.D. thesis, University of Durham, England, 1997.

Chen, Z., Guerrero, J.M. and Blaabjerg, F. (2009), 'A review of the state-of-the-art of power electronics for wind turbines', *IEEE Transactions on Power Electronics*, Vol. **24**, No. 8, pp. 1859–75, August 2009.

Ericsen, T., Khersonsky, Y., Schugart, P. and Steimer, P. (2006), 'PEBB – Power Electronics Building Blocks, from Concept to Reality', The 3rd IET International Conference on Power Electronics, Machines and Drives.

Glinka, M. and Marquardt, R. (2003), 'A new single-phase ac/ac-multilevel converter for traction vehicles operating on ac line voltage', *EPE 2003*, Toulouse, France, September 2–4, 2003.

Habetler, T.G., Profumo, F. Pastorelli, M. and Tolbert, L.M. (1992), 'Direct torque control of induction machines using space vector modulation', *IEEE Transactions on Industry Applications*, Vol. **28**, No. 5, pp. 1045–53.

Helle, L., Larsen, K.B., Jorgensen, A.H., Munk-Nielsen, S. and Blaabjerg, F. (2004), 'Evaluation of modulation schemes for three-phase to three-phase matrix converters', *IEEE Transactions on Industrial Electronics*, Vol. **51**, No. 1, pp. 158–71, February 2004.

Hui Zhang, and Tolbert, L.M. (2011), 'Efficiency impact of silicon carbide power electronics for modern wind turbine full scale frequency converter', *IEEE Transactions on Industrial Electronics*, Vol. **58**, No. 1, pp. 21–8.

Krug, D., Bernet, S. and Dieckerhoff, S. (2003), 'Comparison of state-of-the-art voltage source converter topologies for medium voltage applications', 38th IAS Annual Meeting. Conference Record of the Industry Applications Conference, 2003, Vol. **1**, pp. 168–75.

Marquardt, R. and Lesnicar, A. (2003), 'A new modular voltage source inverter topology', *EPE 2003*, Toulouse, France, September 2–4, 2003.

Noguchi, T., Tomiki, H., Kondo, S. and Takahashi, I. (1998), 'Direct power control of PWM converter without power-source voltage sensors', *IEEE Transactions on Industry Applications*, Vol. **34**, No. 3, pp. 473–9.

Serpa, L.A., Kolar, J.W., Ponnaluri, S. and Barbosa, P.M. (2005), 'A modified direct power control strategy allowing the connection of three-phase inverter to the grid through LCL filters', Conference Record of the 2005 Industry Applications Conference, 2005. Fortieth IAS Annual Meeting, Vol. **1**, pp. 565–71.

Supatti, U. and Peng, F.Z. (2009), 'Z-source inverter with grid connected for wind power system', IEEE Energy Conversion Congress and Exposition, p. 398, 20–24 September 2009, San Jose, CA.

Tolbert, L.M. and Peng, F.Z. (2000), 'Multilevel converters as a utility interface for renewable energy systems', IEEE Power Engineering Society Summer Meeting, PESS.2000, Vol. **2**. pp. 1271–4.

Zhi, D., Xu, L. and Williams, B.W. (2009), 'Improved direct power control of grid-connected DC/AC converters', *IEEE Transactions on Power Electronics*, Vol. **24**, No. 5, pp. 1280–92.

5

Power electronic converter systems for direct drive renewable energy applications

Z. CHEN, Aalborg University, Denmark

DOI: 10.1533/9780857097491.1.106

Abstract: This chapter presents power electronic conversion systems for wind and marine energy generation applications, in particular, direct drive generator energy conversion systems. Various topologies are presented and system design optimization and reliability are briefly discussed.

Key words: direct drive wind and marine power generator, power electronic interface, design optimization, reliability.

5.1 Introduction

This chapter describes power electronic conversion systems for direct drive wind and marine generators. The main function of a power electronic system in a wind or marine power conversion system is to process the electrical power from the generator and send the power into the grid with the required specifications (voltage, frequency and quality). On the generator side, the ideal conversion system would enable the optimal energy to be captured, reducing the system power loss and stress; on the grid side, it would convert the power into the required frequency and voltage with the desired waveform. Furthermore, the power electronic system should be coordinated with other parts of the energy conversion system to perform the desired power control under a variety of conditions.

With the rapid development of renewable energy technologies, power electronics are increasingly used for interfacing renewable generation units into power systems. For direct drive wind and marine power generators, the power electronic system is critical in determining system performance. Figure 5.1 illustrates a direct drive wind turbine system with a full-rated power electronic system consisting of two power electronic converters. The generator side converter converts the variable frequency and voltage AC power from the variable-speed generator into a DC voltage, then the grid side converter converts the DC power into a grid frequency AC power with the desired power quality, and sends it into the grid. The main control

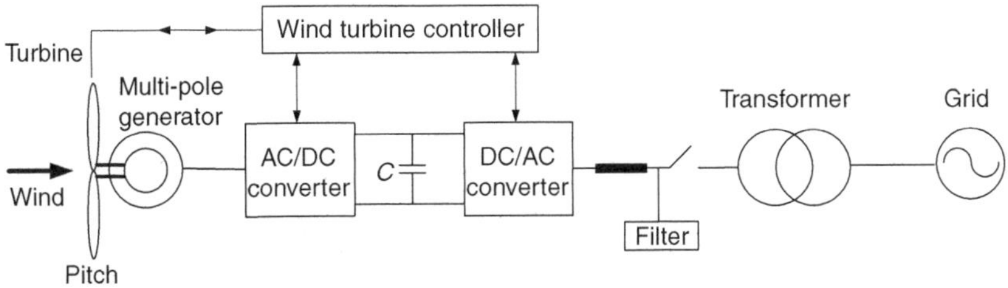

5.1 A direct drive wind turbine systems with a full rate power electronic interface.

functions are performed by the pitch control of the blades and control of the grid and turbine side converters.

However, various converter systems with different topologies and characteristics exist to perform similar functions. Research and development are still very active, with more topologies and control concepts being proposed to further improve system performance.

This chapter is organized into the following sections: Section 5.2 introduces wind and marine energy generation system characteristics. Then the main possible power electronic systems for direct drive wind and marine energy systems are presented, including the back-to-back Voltage Source Converters in Section 5.3; the diode rectifier plus DC/DC converter as the generator side converter in an AC–DC–AC power electronic conversion system in Section 5.4; and the thyristor inverter with active compensator as the grid side converter in an AC–DC–AC power electronic conversion system in Section 5.5. Section 5.6 discusses power electronic system design considerations, and Section 5.7 describes power electronic system challenges and reliability. Future trends are considered in Section 5.8.

5.2 Characteristics of wind and marine energy generation systems

This section briefly describes wind and marine energy generation characteristics, including wind turbines, tidal power conversion system and wave power converters.

5.2.1 Wind turbine power conversion

Wind power is converted into mechanical power by a wind turbine rotor, which drives a generator to produce electricity. The most common form of wind turbine consists of a three-bladed horizontal axis wind turbine, in which the relationship between the wind speed and turbine mechanical power may be described by the following equation:

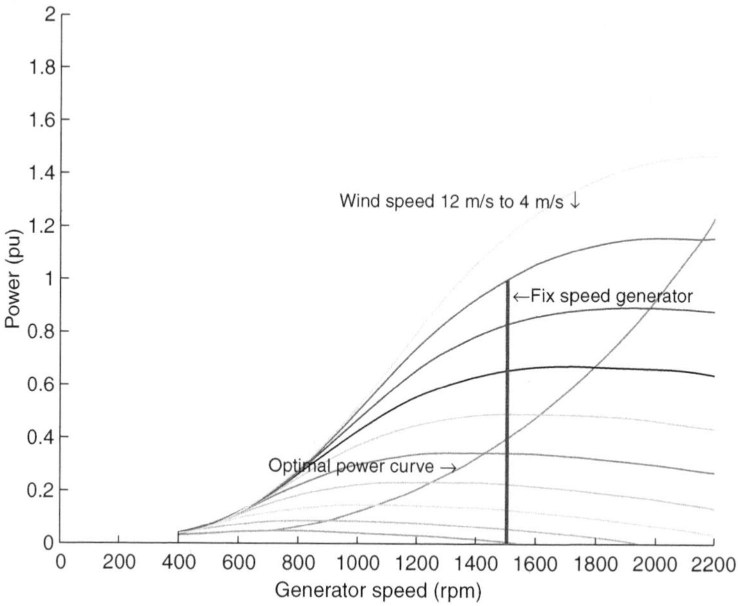

5.2 An example of wind power converter characteristics.

$$P_{\text{wind}} = 1/2\,\rho_{\text{air}}\,\pi R^2 v_{\text{wind}}^3 C_{p.\text{wind}}(\theta, \lambda) \tag{5.1}$$

where

P_{wind} = the power extracted from the wind [W],
ρ_{air} = the air density [kg/m^3], (under normal pressure, 1.225 kg/m^3)
R = the wind turbine rotor radius [m],
v_{wind} = the equivalent wind speed [m/s],
$C_{p.\text{wind}}$ = the wind turbine aerodynamic efficiency coefficient. θ is the pitch angle of wind turbine rotor [deg] and $\lambda = \omega_{\text{rot}} R/v_{\text{wind}}$ is the tip speed ratio, where ω_{rot} is the rotational speed [rad/s].

Normally, a modern wind turbine is controlled to operate at a varying speed to capture the maximum power if the wind speed is below the rated wind speed (as illustrated in Fig. 5.2); and the power is limited by pitch angle or stall if the wind speed is over the rated wind speed. It is important to be able to control or limit the mechanical power into the system to ensure the generator does not exceed the rated power or speed. The common power limitation methods are stall regulation, pitch control and active stall. Because the turbine operates at a variable speed, the generator produces variable-frequency power if an AC generator without a variable ratio gear is used.

5.2.2 Tidal power conversion

Tidal energy is extracted from the relative motion of water. Water levels and tidal currents periodically change because of the gravitational attraction of the sun and moon. The tidal range at a given location is also influenced by the relative positions of the earth, the moon and the sun.

Tidal power may be considered as two components: potential power, $P_{\text{tidal.}p}$, and kinetic power, $P_{\text{tidal.}k}$, potential power is created as a result of the difference in height between high and low tides. $P_{\text{tidal.}k}$ is produced by moving water, which is similar to the principle behind wind turbines.

There are three ways of capturing tidal energy:

1. Tidal stream generator.
2. Tidal barrage.
3. Dynamic tidal power.

Tidal stream generators (TSGs) are underwater tidal turbines which produce mechanical power by converting the kinetic energy from water currents (the kinetic power component), in a similar way to wind turbines which draw energy from air currents. An example is the 1.2 MW SeaGen tidal energy system in Strangford Narrows in Northern Ireland. The TSG method has the advantage of lower cost and lower ecological impact compared to tidal barrages and dynamic tidal power (DTP).

Tidal barrages exploit the potential energy in the height difference between high and low tides. A dam is built to hold the water when the sea level rises, then the water is released to drive generators when the tide is receding. An example of tidal barrage application is La Rance Tidal Power Plant in Brittany, France. The system has a power capacity of 240 MW with an average power production of 68 MW.

DTP is an untried concept. It requires a very long T-shaped dam (30–50 km) from the coast straight out into the sea to exploit both the potential and kinetic energies in tidal flows. No area is enclosed.

The energy available from a TSG can be expressed as:

$$P_{\text{tidal.}c} = 1/2\, \rho_{\text{water}} A v_{\text{water}}^3 C_{p,\text{tidal}} \qquad [5.2]$$

where

$P_{\text{tidal.}c}$ = the power generated (W)

ρ_{water} = the density of the water (seawater is 1027 kg/m^3)

A = the swept area of the turbine (m^2)
v_{water} = the velocity of the flow (m/s)
$C_{p.tidal}$ = the turbine power coefficient

The higher density of water relative to air means that a single TSG can provide significant power at low tidal flow velocities. Tidal generators have operational characteristics very similar to those of a wind turbine; therefore, variable-frequency power is normally generated from an AC generator.

5.2.3 Wave power conversion

Waves are generated by wind passing over the surface of the sea. Wave power generators capture the energy of the waves and are generally categorized according to the method used, including point absorber or buoy; overtopping; oscillating water column; terminator located perpendicular to the wave motion; and surfacing attenuator parallel to the wave motion. Wave power devices can also be categorized by location, such as shoreline, near-shore and offshore.

Wave power devices traditionally use conventional rotary electrical machines for power conversion, where hydraulic systems or air turbines are required to convert the low reciprocating motion of the wave device to rotation at a speed suitable for normal generators, such as 1500 rpm. In a direct drive system, a reciprocating electrical machine is driven at the same speed as the mechanical device. Linear permanent magnet synchronous machines have been investigated for this application and show potential (Mueller, 2002; Polinder *et al.*, 2004; Leijon *et al.*, 2005). The fluctuating power generated can then be conditioned by a full-rated power electronic system for grid transmission and customer consumption.

Direct drive wind or marine energy conversion systems have some common features: the generator is designed and operated to optimize the energy conversion, and normally produces fluctuating power with varying frequency and voltage. Consequently a full-rated power electronic converter system is needed to decouple the generator and grid, to convert a variable-frequency AC voltage from the variable-speed generator to a voltage with the fixed frequency of the AC grid. Because of this similarity, similar converter configurations may be used for direct drive wind and marine energy systems. However, as wind power is a much more mature energy technology and the associated power electronic systems are well developed, the main focus of this chapter will be power electronic converters for wind power. Such converters can be easily adopted for marine energy systems. Although direct AC–AC conversion is possible, most commonly used converter systems have a basic AC–DC–AC configuration. Various grid interface options for a variable-speed, permanent magnet modular wind power generator have been discussed by Chen (1997) and Chen and Spooner (1998).

5.3 Back-to-back voltage source converter (BTB-VSC)

Figure 5.3 shows a back-to-back pulse width modulation (PWM) voltage source converter system, which is the most commonly used power electronic interface system in wind power conversion. The power electronic system consists of two bidirectional PWM voltage source converters connected back to back (BTB). One is on the generator side and the other on the grid side. The two converters share a common DC link supported by a DC capacitor.

A PWM VSC has the ability to control both the active and reactive power at its AC terminals. The BTB configuration gives the potential for both optimizing the generator operation and satisfying the grid integration requirements.

The reactive power of the two converters is independent, and can therefore be controlled separately. For example, the reactive power at the generator side can be controlled to minimize the power loss of the generator or to keep a constant generator terminal voltage; the reactive power of the grid side converter can be controlled as zero (unit power factor) or at values desired for good power quality or a value required by the system operator within the converter rating limit. However, the active power controls of the two converters are related. Ignoring the power loss of the DC link, the active power from the generator side should be equal to the active power delivered into the grid under stable operating conditions so that a constant DC-link voltage is maintained. Consequently, control of the two converters' active power needs to be coordinated to keep the DC-link voltage within the specified range.

The commonly used vector controller described in Section 4.5 normally has a nested-loop structure: the fast inner current loops control the dq-axis

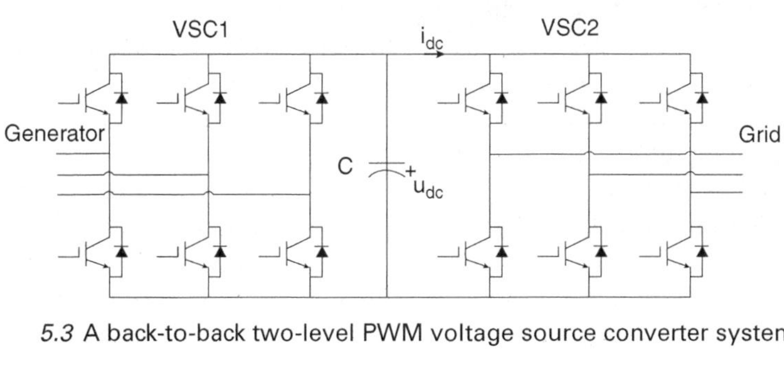

5.3 A back-to-back two-level PWM voltage source converter system.

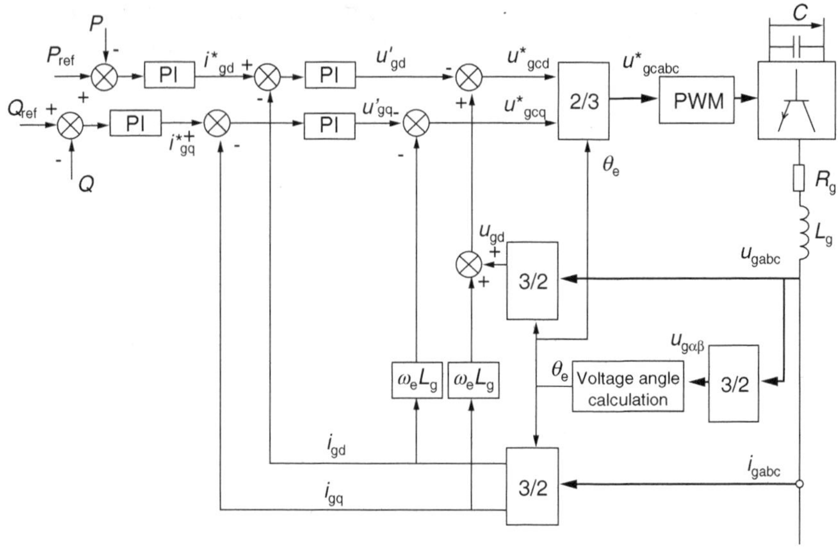

5.4 *P, Q* control of a vector-controlled PWM VSC (AC voltage aligned on d-axis).

currents, and the outer slower loops are used for torque/power and reactive power control.

Figure 5.4 shows a system based on a voltage-oriented controller where the *d*-component current reference is generated from an active power reference, though the current reference could also be obtained from torque or speed reference, while the *q*-current component reference is set to follow a reactive power reference.

If the DC-link voltage in Fig. 5.3 is to be kept at a constant level, then all the active power entering from the generator side will be sent into the grid side by the grid side converter (assuming the DC-link power losses can be ignored). Therefore, the DC-link voltage can be used as a control variable for one of the PWM VSCs. Normally, one converter performs the power control by following a specified power reference (such as shown in Fig. 5.4) and the other converter regulates the DC-link voltage, as shown in Fig. 5.5, where the converter uses the i_{sd} demand derived from the DC-link voltage error.

Depending on the tasks of the converters, two control schemes may be arranged as shown in Table 5.1. The two converters' functions of controlling active power or regulating the DC-link voltage may be exchanged as shown in Table 5.1. The role exchange may make no difference in normal operation. However, in the event of a grid fault, the grid side converter may not be able to function appropriately, then Control scheme II is more effective in terms of DC-link voltage control (Hu *et al.*, 2009a).

Table 5.1 Control schemes of BTB-VSC PWM converters

	Control scheme	Generator side converter (PWM converter)	Grid side converter (PWM converter)
Active power	I	Following a reference active power/torque	Regulate the DC-link voltage
	II	Regulate the DC-link voltage	Following a reference active power/torque
Reactive power	I and II	Control generator terminal voltage or power factor or reactive power	Control reactive power sent into the grid or power factor or grid voltage

Figure 5.6 shows some schemes of deriving the reference signals for the power electronic control systems shown in Figs 5.4 and 5.5. Figure 5.6a shows that a reference active power is determined by the machine speed, which is normally used for realizing the maximum power capture (Chen, 1997; Chen and Spooner, 1998). In Fig. 5.6b a reactive current reference is chosen to keep the desired stator terminal voltage, while Fig. 5.6c shows a control scheme to damp the active power fluctuation by using a varying DC voltage (Michalke *et al.*, 2007). The system in Fig. 5.6d minimizes flicker by varying the DC-link voltage to smooth the power fluctuations from a wind power generator (Hu *et al.*, 2009b).

The basic control principles of all the voltage DC-link type converters described in Section 4.3 can be used in the BTB-VSC configuration for the

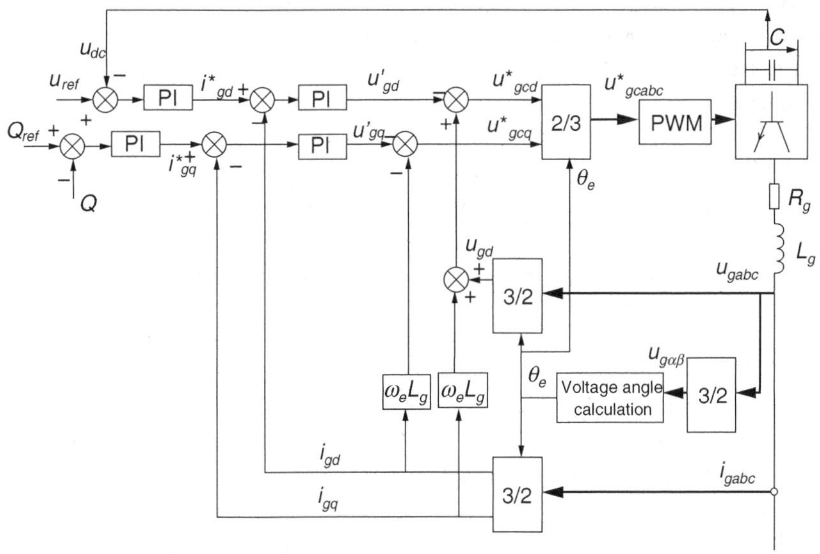

5.5 u_{dc}, Q control of a vector-controlled PWM VSC (AC voltage aligned on d-axis).

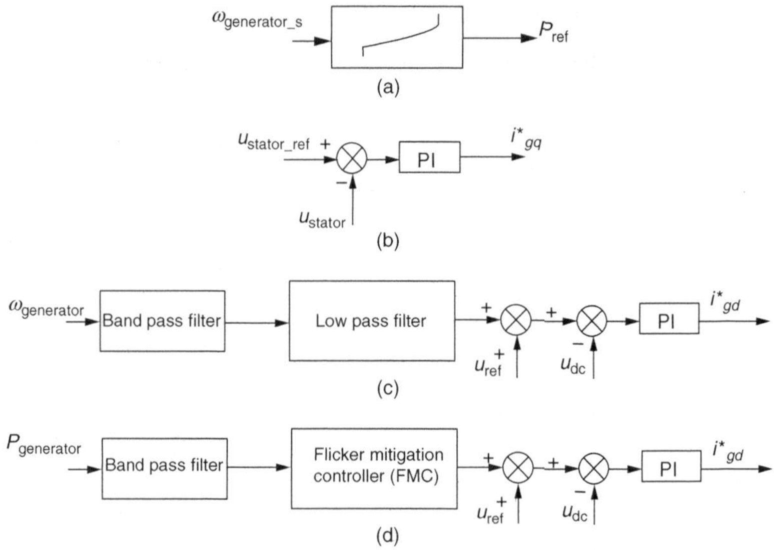

5.6 Some schemes of generating reference variables of u_{dc-ref}, P_{ref}, Q_{ref} for a *dq* decoupled vector-controlled voltage source converter. (a) Active power reference generated from the wind turbine speed; (b) reactive current reference for generator terminal voltage control (reactive current at q-axis); (c) damping of active power fluctuation with varying DC voltage (grid converter active current reference); and (d) flicker mitigation with varying DC voltage (grid converter active current reference).

relevant renewable energy application, though some variations on the converter modulation strategies may be needed.

The following sections describe some example configurations (Blaabjerg *et al.*, 2010).

5.3.1 Three-level neutral-point diode-clamped back-to-back topology (3L NPC-BTB)

A three-level neutral-point diode-clamped back-to-back (3L NPC-BTB) converter is shown in Fig. 5.7. The 3L-BTB has more output voltage levels and less dv/dt stresses than the 2L-BTB, thus the filter can be smaller. The mid-point voltage fluctuation can be dealt with by controlling the status of redundant switches. However, the power loss may be unequally distributed between the outer and inner devices in a leg which could lead to de-rated converter power capacity (Krug *et al.*, 2007).

A converter project for a 5 MW offshore wind turbine was reported (Eichler *et al.*, 2008), in which the converter has two NPC (neutral-point connected) converters based on high-power semiconductor IGCT–PEBB (modular power electronics building blocks) technology, leading to the development of compact converters.

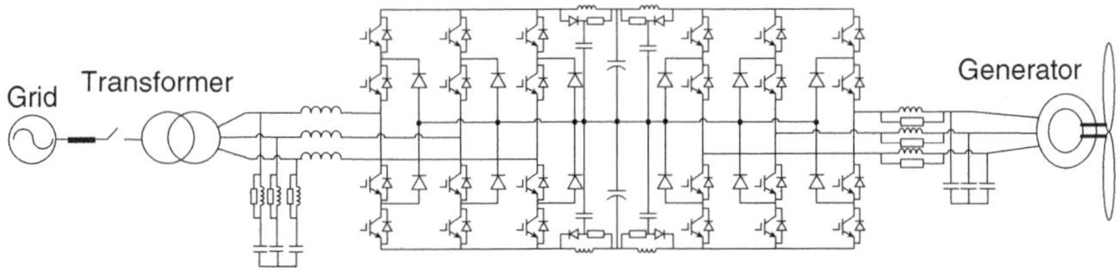

5.7 Three-level Neutral-Point Clamped back-to-back converter (3LNPC-BTB).

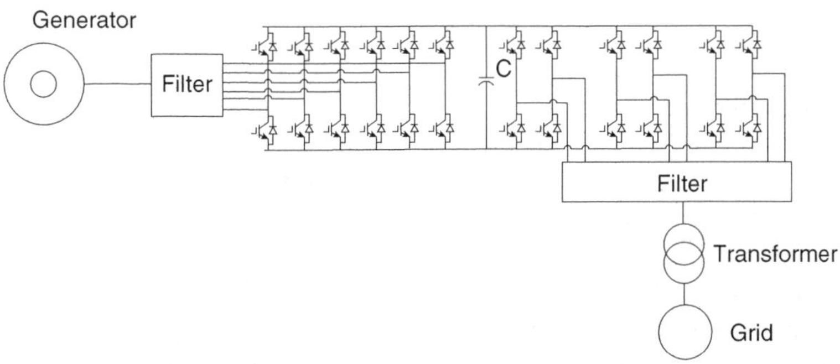

5.8 Three-phase H-bridge back-to-back converter (3P HB-BTB).

5.3.2 Three-phase H-bridge back-to-back topology (3P HB-BTB)

As shown in Fig. 5.8, each phase of the 3P HB-BTB configuration consists of two H-bridge converters located at the grid and generator sides, respectively. It can achieve a similar output to the 3L NPC-BTB circuit while eliminating the unequal power loss distribution and clamped diodes. More efficient and equal usage of switching devices and higher designed power capacity could be achieved (Senturk *et al.*, 2009).

However, a 3P HB-BTB requires separate single-phase accessible configuration in the generator and grid side transformer in order to achieve isolation between each phase, which results in more cable connections and heavier weight, and consequently, extra cost and loss, though this feature has fault tolerance potential.

5.3.3 Combination of a multi-level topology and a multi-H bridge topology

Generally, the power quality requirements on the grid side will be more stringent than on the generator side (Krug *et al.*, 2007). To meet such features, various of the circuit topologies discussed may be used in combinations, for example, a multi-level circuit on the generator side, and a multi-H bridge topology on the grid side to achieve the required performance cost-effectively.

5.3.4 Cascaded H-bridge back-to-back converter with medium-frequency transformers (CHB-MFT)

The cascaded H-Bridge (CHB) converters described in Section 4.3 need an isolated DC-link for each converter cell. This may involve a complex multi-pulse transformer, resulting in larger weight and volume. The circuit configuration shown in Fig. 5.9 (Iov *et al.*, 2010) is based on a back-to-back cascaded H-bridge converter structure, with galvanic insulated DC/DC converters as interface.

A DC/DC converter with medium-frequency transformer (MFT) may operate from several kHz to dozens of kHz, thus reducing the transformer size. Because of the cascaded structure, this configuration may be directly connected to the renewable power collection network at a distribution voltage level (10–20 kV) with high output voltage quality, filterless design, and redundancy (Inoue and Akagi, 2007).

However, the configurations using a large number of power semiconductors as well as auxiliary components could reduce this converter's reliability and increase the cost.

5.4 Diode rectifier plus DC/DC converter as the generator side converter

Direct drive generators are normally low-speed synchronous generators, either electromagnetically excited or permanent magnet (PM) excited. The converter on the generator side does not have to provide excitation for the synchronous generator, therefore, a diode rectifier may be used for simplicity and low cost (Chen and Spooner, 1995, 1998, 2001a; Chen, 1997).

The variable-speed PM generator outputs a voltage varying in frequency and magnitude which could be rectified by a diode rectifier to form a variable-voltage DC link. The limitation on the power transfer due to the internal inductance of the generator can be overcome by shunt-connected AC capacitors (Chen *et al.*, 1998), Such a method has also been used in direct drive wave generator investigation (Mueller, 2002). A Forced Commutation Controlled Series Capacitor (FCSC) technique (Pickert *et al.*, 2011) has been discussed to improve the power factor of the variable amplitude and low and variable-frequency voltage produced by linear wave generators.

A DC/DC converter may be placed between the generator-rectifier and the grid side VSC as shown in Fig. 5.10 (Chen, 1997; Chen and Spooner, 2001a).

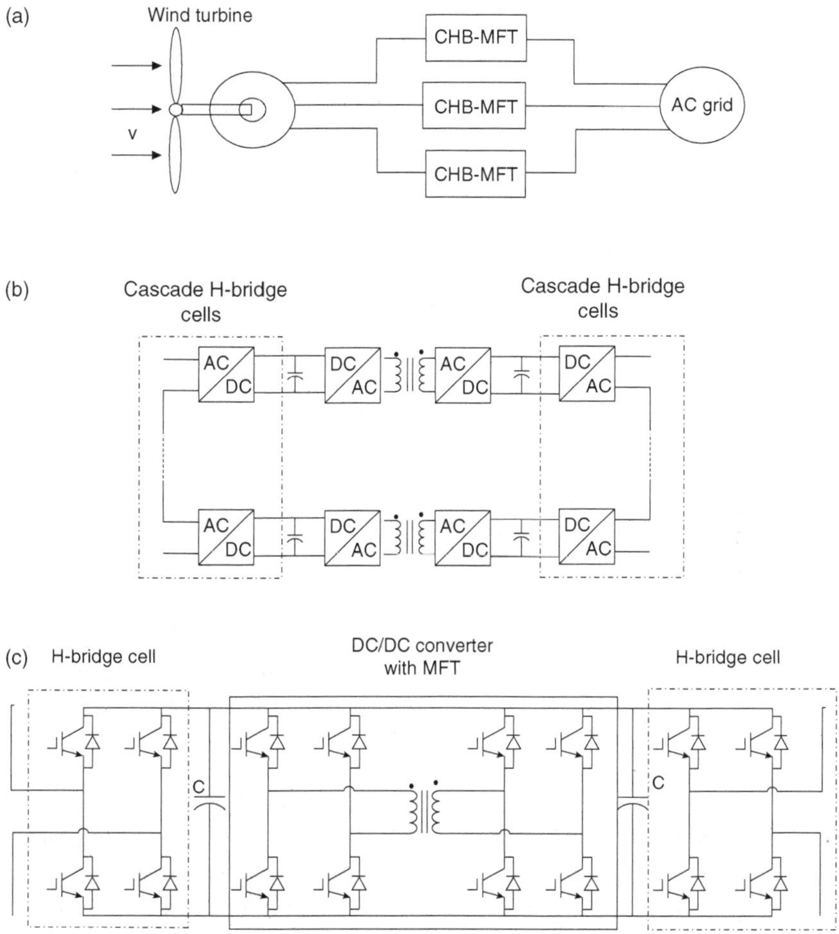

5.9 Cascade H-bridge back-to-back converter with medium-frequency transformer (CHB-MFT). (a) Three phase system illustration; (b) one phase configuration and (c) H-bridge back-to-back converter with medium -frequency transformer.

The DC/DC converter separates the DC link into two voltage levels, one at the rectifier terminal for active power control of the generator-rectifier and the other at the inverter terminal to follow a specified voltage curve or simply a constant DC voltage. The modulation ratio of the grid side VSC can then be fixed or varied over a greatly reduced range for high utilization of the converter (Chen and Spooner, 2001a). The configuration and control of the grid side converter can be the same as discussed above (see Section 5.3). The effect of such a converter on grid harmonic distortion distribution has also been analyzed (Chen and Spooner, 2001b), and it has been shown that the appropriate filtering equipment may have to be used to meet the grid harmonic standard.

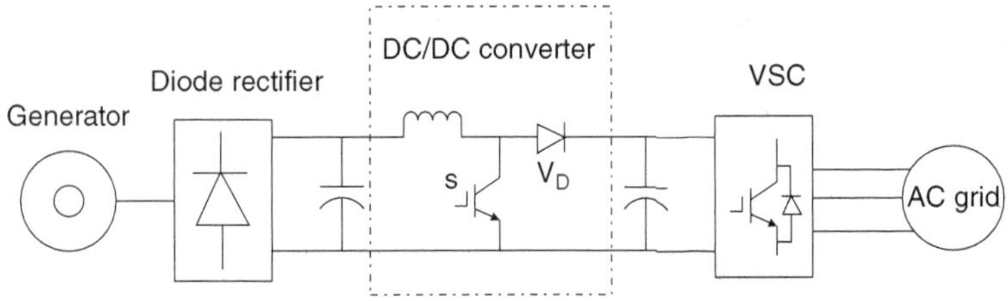

5.10 Power electronic conversion system with diode rectifier, DC/DC converter-VSC (Chen, 1997).

Table 5.2 Control schemes of Diode rectifier + DC/DC converter and PWM converter

	Control scheme	Generator side converter (Diode rectifier + DC/DC converter)	Grid side converter (PWM converter)
Active power	I	Following a reference active power/torque	Regulate the DC-link voltage
	II	Regulate the DC-link voltage	Following a reference active power/torque
Reactive power	I and II	No control	Control reactive power sent into the grid or power factor, or grid voltage

The control tasks described for BTB-VSC could also be performed in a similar way to those for back-to-back VSC PWM converters, except that the reactive power cannot be controlled with the diode rectifier on the generator side. Possible control schemes are summarized in Table 5.2. Similarly, the active power-related control functions of the two converters may be exchanged with no difference in normal operation, and in the event of a grid fault, Control scheme II may have better controllability for DC-link voltage regulation (Deng and Chen, 2009).

5.4.1 Multi-modular diode rectifier systems

Depending on the generator configuration, the power electronic rectifier system could have different configurations. For a modular PM generator (Chen and Spooner, 1995), the diode rectifier configuration shown in Fig. 5.11 may be used (Chen, 1997). Connecting each coil of the generator stator modules individually to a star or an H-bridge rectifier makes the complex interconnections of a three-phase winding unnecessary. The outputs of each coil and rectifier may be connected in parallel to the DC link. The DC-link voltage could have very little ripple because of the large number of diode H-bridges.

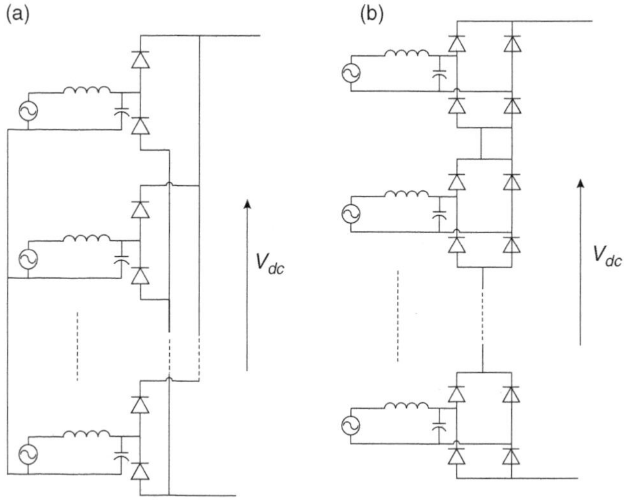

5.11 Diode rectifier configurations for modular PM generators (Chen, 1997). (a) Star stator coil connection and (b) H-bridge stator coil connection.

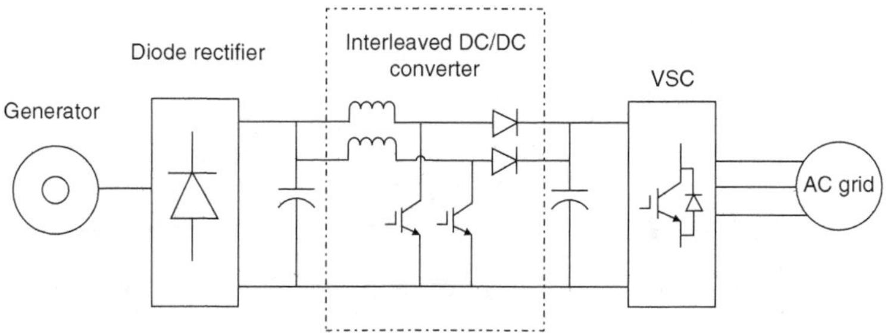

5.12 Power electronic system with an interleaved DC/DC converter.

5.4.2 Interleaved DC/DC converter topologies

For DC/DC converters, an interleaved configuration such as the system illustrated in Fig. 5.12 may also be used to provide higher power capacity and better power quality (Li *et al.*, 2009).

5.4.3 Multi-voltage source converter topologies

Various converter topologies can be used for the grid side converter, such as multi-level converters and multi-H bridge converters, etc. Furthermore, those converters may be connected into multi-converter topologies for higher power delivery and good power quality.

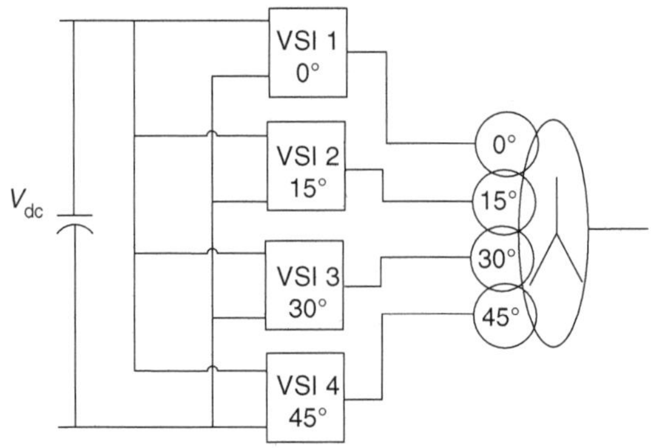

5.13 A 24-pulse inverter with a multi-winding transformer (Chen, 1997).

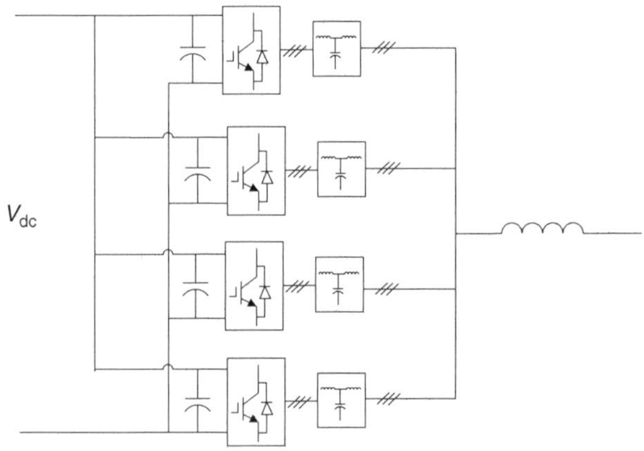

5.14 Multi-interleaved converter power electronic grid interface.

Figure 5.13 shows a 24-pulse voltage source converter system with a multi-winding transformer (Chen, 1997), where four 6-pulse voltage source converters are connected in parallel to the same DC link, and the converter AC voltage outputs are phase-shifted by intervals of 0°, 15°, 30° and 45°, respectively and connected to a transformer which has multiple windings and in turn advances these converter voltages by 0°, 15°, 30° and 45°, respectively to synthesize a 24-step voltage waveform.

Figure 5.14 shows a power electronic grid interface with a multi-interleaved converter system. The system is different from that shown in Fig. 5.13, in that the special transformer is not required.

A study on the impact of interleaving on harmonic current in DC passive components has been reported (Zhang *et al.*, 2008). The effects of various factors have been analyzed, including modulation index, displacement angle and the interleaving angle.

A modulation and control method for parallel-connected inverters was investigated in the EU-FP6 UPWIND project (Hemmelmann, 2008); a proposed multi-level space vector modulator can reduce the requirements on the differential and the common mode choke. The cross-current could be limited by a balancing choke.

For large wind turbines, the power electronic system could be constructed with either a higher-voltage power electronic converter or a number of lower-voltage power electronic converters. A comparative study has been conducted for a 6 MW wind turbine converter (Troedson and Desai, 2010). The results show that the multi-low voltage converter is still a cheaper option, even though a medium-voltage converter system may be more suitable for larger wind turbines.

5.5 Application of current source converters (CSCs)

Although VSCs are commonly used for wind power conversion system, the current type of DC link can also be used with current source converters (CSCs). In this section, two types of CSCs are briefly discussed, the thyristor converter with active compensator as grid side converter, and PWM CSC.

5.5.1 Diode rectifier plus silicon controlled rectifier current source converter (SCR-CSC) at grid side

The application of thyristor or silicon controlled rectifier (SCR) current source converter (SCR-CSC) in the wind power conversion system has also been discussed (Chen, 1997; Chen and Spooner, 2003). Thyristors have high-power capacity, but normally operate at low switching frequency in line commutation mode, and require reactive power compensation and harmonic filtering. If such a thyristor converter is used over a wide range of DC voltages, as in the case of interfacing a variable-speed direct drive permanent magnet generator in wind power systems, the converter may be required to operate over a wide range of firing angles to perform the active power transfer. Consequently, reactive power and harmonic problems could become more significant and could require compensation. Various compensation options could be used, including the passive, active or hybrid compensation schemes studied (Chen and Spooner, 2003; Chen, 2004). One such systems is shown in Fig. 5.15.

Multi-pulse converter topologies, like voltage source converters, can be used to improve the power quality and deal with higher power (Chen, 1997; Chen and Spooner, 2003). A 12-pulse SCR-CSC topology is shown in Fig. 5.16.

An SCR-CSC needs a strong AC system or other facilities to ensure commutation, whereas a self-commutated VSC can operate independently. Therefore, a SCR-CSC system may be used for a strong AC system, while a self-commutated VSC is suitable for a weak or isolated system.

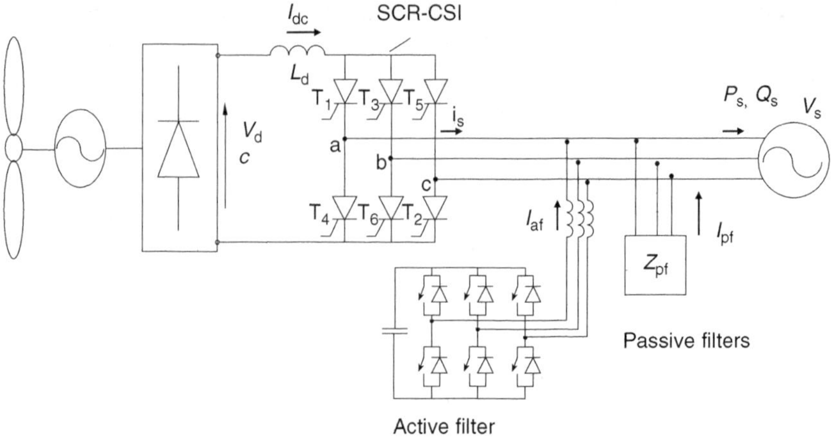

5.15 SCR-CSC grid converter with parallel-connected hybrid compensators (Chen and Spooner, 2003).

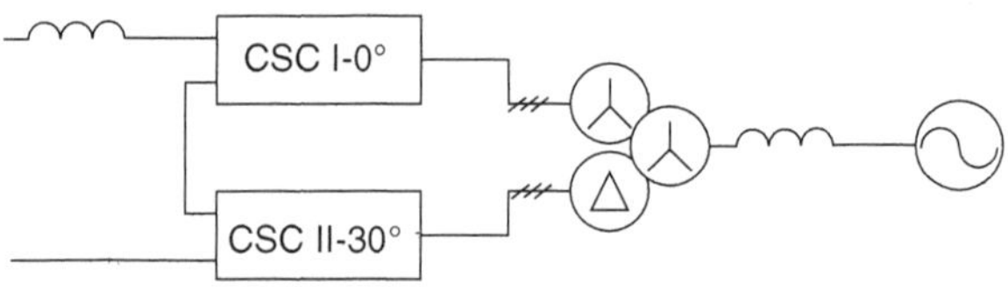

5.16 A 12-pulse SCR-CSC (Chen, 1997).

While self-commutated semiconductor-based voltage source converters can be switched off very rapidly as a means of circuit protection without any difficulty arising from stored energy, an SCR-CSC cannot be turned off by gate control, and cutting off the DC current would generate large overvoltage. Also, during a power system fault, the self-commutated VSC (main inverter or active compensator) may lend support with voltage regulation and reactive power control abilities.

5.5.2 Pulse width modulation (PWM) current source converters

Current source converters can also be implemented with controllable semiconductors, so that a PWM current waveform can be produced at the converter AC terminals. A converter system with diode rectifier at the generator side and a PWM-CSC implemented with fully controllable semiconductors on the grid side was studied by Tenca *et al.* (2007). Furthermore, a back-to-back PWM current-source converter (CSC) topology is proposed for high-power wind energy applications by Dai *et al.* (2009). Compared

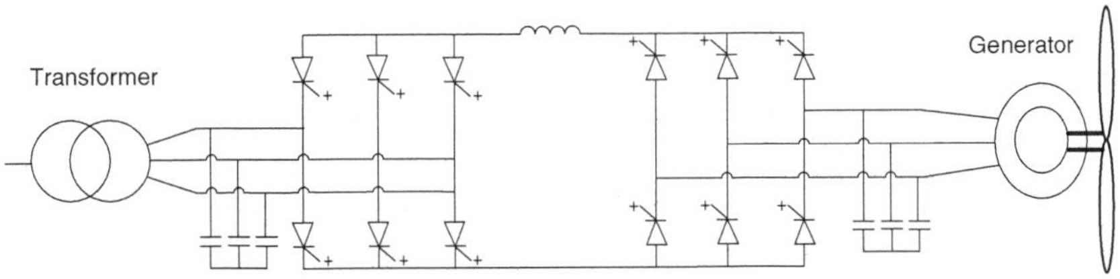

5.17 Configuration of power electronic conversion system with BTB PWM-CSC.

with VSC and LCI-based configurations, PWM-CSCs have a simple topology and good grid integration performance. In addition the DC-link reactor provides natural protection against a short-circuit fault. A back-to-back PWM-CSC conversion system is shown in Fig. 5.17.

5.6 Power electronic system design considerations

Designing a power electronic system involves a large number of design variables in different engineering fields, covering electrical, magnetic, thermal and mechanical aspects (Busquets-monge *et al.*, 2004). Usually some simplifications have to be made. Mathematical optimization techniques offer a systematic way of handling the design problem. This allows the designer to use more design variables and fewer simplifications. The increasing computation capacity of modern computers and the development of efficient computational algorithms allow the designs to be optimized in a relatively short time. Furthermore, the application of the optimization techniques can provide a better understanding of the trade-offs involved in the design.

A design optimization of a power converter involves selecting a set of design variables under a set of constraints, to optimize certain converter characteristics, such as the converter cost, weight, power loss, or any other physically realizable characteristic.

The component ratings, switching frequency, operational voltage and associated filter parameters are usually considered as design variables, which are discrete in nature. The component parameters may be obtained directly from the data sheets and could be experimentally calibrated and modified if necessary. The databases of commercially available components are associated with the necessary parameters in electromagnetic, thermal, mechanical, reliability and cost aspects.

The design constraints are defined according to the application specifications and physical limitations, such as power delivery capacity, power quality standard and component safe operating areas. One of the discrete optimization methodologies, such as a genetic algorithm (GA), can be used for power electronics system optimization (Busquets-monge *et al.*, 2004).

In general, a power converter design may contain the following major steps.

1. Specify converter requirements, such as power rating of the converter, and select the basic power circuit configuration.
2. Determine the semiconductor and other converter component parameters.
3. Choose the modulation and control methods.
4. Determine the power converter switching frequency and design filters to meet power quality requirement, such as voltage ripple and harmonics.
5. Check the behaviour of the designed power electronic system (such as power delivery and control ability, system power losses, stability, responses to disturbances, etc.), and evaluate the design objective function and other indices of interests with reference to the specifications.

The above procedure or a part of the procedure may be repeated for different sets of design variables to optimize the design, for example, different switching frequencies and associated filters. Simulation methods can be used to check the designs, and experimental prototypes may be implemented to perform tests with the selected design.

In practice, the design focuses principally on the power circuit. The number of design criteria may be reduced to include only those related to the power circuit. For example, the most important parameters – cost, weight, size or losses – dominate the design of power circuits in higher-power converters. If the power circuit and its parameters are appropriate, it is possible to design a compatible controller to meet control-dependent performance requirements, such as stability. Thus, the control-dependent constraints are unlikely to alter the main results obtained by considering only the power circuit-related constraints (Balachandran and Lee, 1981).

There are often only a few common power circuit options, and they are well defined. A practical approach to power electronic system design, therefore, is first to design the power circuit to optimize cost, weight or loss. The detailed controller design is then based on the power circuit to meet the specified control-dependent performance requirements.

Simulation is a very useful tool for power electronic system design, analysis and evaluation. The following may be simulated:

- **Circuit design**: To obtain current and voltage waveforms of a converter under different operating conditions. All component voltage and current ratings can be checked and the harmonics of the converter system are calculated to design filters to meet power quality and EMI requirements.

- **Power loss evaluation**: To calculate power loss of the components by using the relevant voltage and current waveforms, various operation conditions can be considered.
- **Thermal design/cooling system**: To analyze the semiconductor junction temperatures by using power losses under different operating conditions, transformer and inductor temperatures (including transient temporary overload conditions), and evaluate the performance of various possible cooling systems.

5.6.1 An example of a comparative study

The following comparative power circuit design was studied by Zeng *et al.* (2007). The system is a full-rated power electronic conversion system for a wind power generation system. Three BTB-VSC configurations are considered: two-level, three-level neutral-point clamped, and three-level flying capacitor BTB converters. The following principal aspects are evaluated for cost, power losses and power quality: converter rating, semiconductor rating/DC-link voltage level, filter design, modulation methods and control strategies.

The ratings of the generator side converter and the grid side converter may be decided separately. The characteristics of the generator (power, voltage and current) will determine the rating of the power electronic converter at the generator side, while the grid side converter will be influenced by the active power capacity, grid reactive power control, fault ride-through ability and grid power quality requirements. It may also be affected by other specified grid operation and control strategies.

The DC-link voltage is an important parameter affecting semiconductor ratings, as well as passive components such as DC capacitors. Together with the filter design and grid impedance, the DC-link voltage also determines the power delivery capacity of the converter system. The voltage drop across the filter inductor would require a suitable DC-link voltage to deliver the power into the grid. Space vector pulse width modulation (SVPWM) strategies are used for these converters. The neutral-point potential balance of 3L-NPC VSC and the voltage unbalance on the fly-capacitor of 3L-FLC VSC are treated.

The harmonic standard and power system operation requirements for the converters are considered. The simulation models are developed in Simulink and Matlab, and simulation studies are performed to verify the designs and to analyze power losses of the converters. Some simulation results are shown in Fig. 5.18.

The costs of passive components become significant and sensitive to switching frequency. Three-level (3L) converters show great advantages

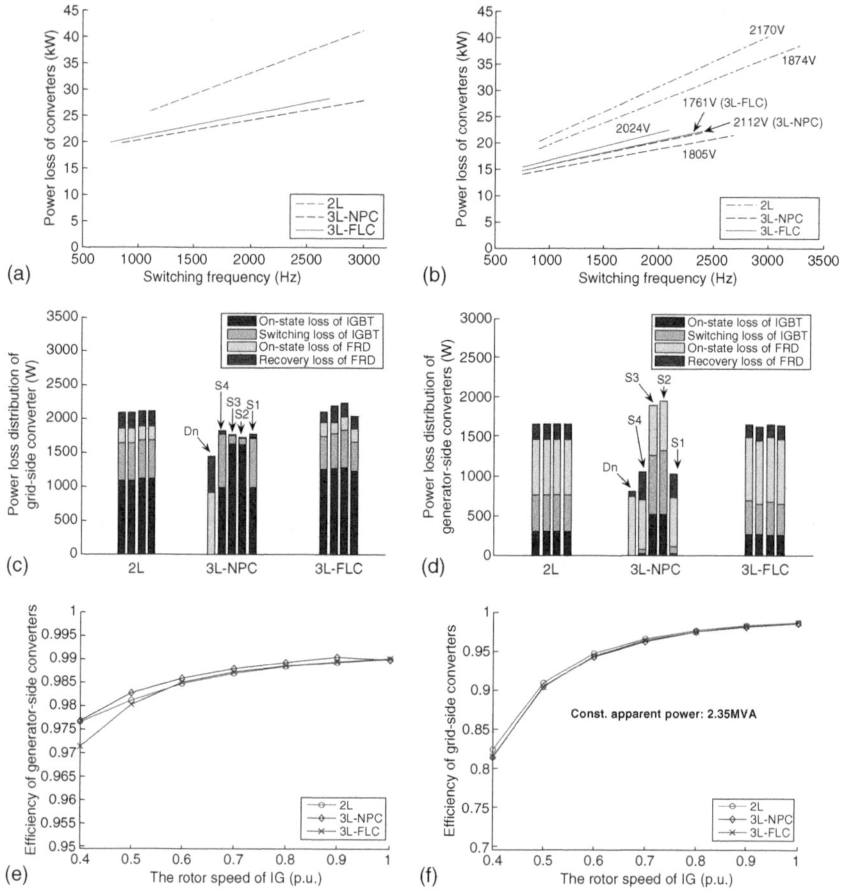

5.18 Power loss simulation results of the converters (Zeng *et al.*, 2007). (a) Grid-sidepower losses *vs.* f_{sw}. (b) Gen-side power losses *vs.* f_{sw}. (c) Grid-side power loss distribution. (d) Gen-side power loss distribution. (e) Gen-side efficiency *vs.* rotor speed. (f) Grid-side efficiency *vs.* rotor speed.

over two-level converters if the same efficiency is required. Better harmonic performance in 3L converters can reduce the size of the filter inductor significantly.

The filter design will be determined by the harmonic requirement, converter switching frequency, modulation methods and power delivery capacity. The filter should be designed to meet the total harmonic distortion (THD) standard under a converter modulation strategy, while the converter switching frequency is a key parameter. A higher switching frequency will reduce the filter components, but increase the power losses. Furthermore, the filter parameter may affect the DC-link voltage required to deliver a certain level of power. Therefore consideration should be given to producing a design that is cost-effective overall.

The costs of active and passive components for the three converters studied are also compared (Zeng *et al.*, 2007). Costs of commercial semiconductor devices and capacitors are provided by the manufacturers or distributors. The material costs of filter inductors are estimated. The costs of the 2L VSC are found to be quite high because of the large LC filters resulting from low switching frequency. In contrast, the 3L VSC shows an improved power loss and harmonic performance, so high switching frequencies and small LC filters may be adopted, greatly reducing costs. In the comparative study, the designs are assumed to have the same cooling system, but the cooling system could be a determining factor particularly in the cost of high-power applications when different voltage levels are considered.

The above example illustrates the general considerations of designing a power electronic system. The procedure may be used as a guide to converter design, though a real design may have to take more practical issues into consideration.

5.7 Power electronic system challenges and reliability

The application areas of power electronics are continuing to grow; meanwhile, the challenges and demands of the technology are also increasing, especially in extreme environments, such as offshore and deep-water situations for offshore renewable power generators. The traditional converter design needs to be improved: for instance, special encapsulation would be required for semiconductors and capacitors to operate successfully in high-pressure conditions, and semiconductor capacity may also be affected by air pressure. System reliability and environment-tolerant technologies are very important, since installation and maintenance costs in such extreme environments will be very high. The desired power electronic system features that are required in offshore and underwater conditions may include efficient operation, reliable components and converters, acceptable costs, and easy accessibility for on-site monitoring and repair.

5.7.1 Factors affecting reliability

Power systems consisting of power electronic converters are more complex than traditional power systems and have some distinctive features. For example, the electrical overload capacity (voltage/current) of power semiconductors is strictly limited. A semiconductor must be switched off within a few microseconds in an over-current situation in order to avoid device breakdown. The following factors affect the reliability of a power electronic converter:

- Reliability and robustness of components (power semiconductors, capacitors, transformers, electronic components, etc.).
- Component derating factors (voltage, current, temperature, etc.).
- Power circuit topology and design.
- Selection and implementation of control strategies (including monitoring, fault handling and protection).

Power semiconductors are critical components when considering power electronic converter reliability. Excessive stress and wear during operation could cause characteristic degradations and potential failures in the long term. A significant percentage of failures in electronic systems are temperature-related. Therefore, reliability analysis methods are associated with both power electronic circuits and thermo-mechanical models. The calculation of module and/or converter system reliability under transient temperatures should be an important feature of power electronics simulation software. The finite element method (FEM) can be used to calculate thermo-mechanical behaviour, for example bond wire lift-off and solder fatigue. Component failure mechanisms include the following (Yang *et al.*, 2010):

- Electrical overstress, such as overvoltage and over-current.
- Electrostatic discharge, such as puncturing the gate oxide.
- Latch-up and triggering of parasitic – excessive dv/dt during turn-off may trigger the parasitic thyristor in IGBTs, causing-latch up.
- Charge effects – ionic contamination (the electric field distortion by the accumulation of ionic contaminants) or hot carrier injection (the growth of defects in the gate oxide).
- Electro-migration, contact and stress-induced migration, which are related to the metallization of the semiconductor devices and affect long-term reliability.
- Thermal activation – the degradation processes of silicon power devices are accelerated by increased temperature.
- External radiation – mobile ions and particles may affect the bipolar and MOS devices by external radiation, and the protons and electrons in the earth's magnetic field may cause ionization and displacement damage in power semiconductor devices.
- Thermo-mechanical fatigue, such as bond wire lift-off and solder fatigue.

For power electronic devices, thermal fatigue unreliability induced by power cycling is an important concern. Since power cycling produces non-uniform temperature distribution in the device, a coupled thermal structural analysis is required to evaluate the thermal fatigue mechanism. The finite element method (FEM) can be used to evaluate the temperature distribution induced by joule heating (Anzawa *et al.*, 2007).

The thermo-mechanical fatigue stress experienced by packaging materials is caused by the mismatch in the thermal expansion coefficients of different materials in the device and package, and the local temperature variations. Thermo-mechanical fatigue is the dominant wear-out mechanism in modern IGBT-based power converters. The reliability record for press-pack devices is much higher than for power modules, largely due to a better tolerance of thermal cycling: the press-pack allows double-sided cooling to lower the thermal impedance.

A survey of the reliability of power electronic systems found that power electronic device failures contribute to 31% of the total faults, while capacitors are 18%, gate drivers 14%, connectors 12%, inductors 5%, and resistor 2% (Tavner, 2012).

5.7.2 Reliability evaluation

Power electronic system reliability analysis can help to get an understanding of the root cause of failures, downtime, and cost, and develop methods to effectively improve reliability. One direct approach is to collect field data and use it to calculate reliability and the relationship to each cause/component. Another approach is to model the functional structure of the system using a fault tree to derive system reliability from component reliability.

Sampling methods used to analyze the reliability include Monte Carlo analysis, importance sampling, first-order reliability method (FORM), and second-order reliability method (SORM) (Kamas, and Sanders, 1996).

Monte Carlo analysis is a statistical sampling method which employs a random selection of parameter values in such a way that they reflect the random physical processes of the original problem. It is generally believed to be the most accurate for estimation. The application of the Monte Carlo method to analyze the reliability of power electronic systems allows the consideration of a random variation of electric load. The probability of failure, *Pf*, can be estimated from computations based on the random parameter values. However, the Monte Carlo method requires intensive computation and it may be prohibitive for cases where a large number of simulations is required.

Importance sampling is a simplified version of the Monte Carlo method. It reduces the variance in the approximation for a given number of samples and can also be used to reduce the number of samples needed for a given variance.

First-order reliability method (FORM) conducts simulations in a gradient-based search for the statistically most likely point of failure for a particular performance specification under a given design centre and parameter distributions.

Second-order reliability method (SORM) approximates the limit-state surface near the most likely point of failure with a curvature model.

5.7.3 Fault tree for reliability analysis

Fault-tree analysis was first proposed in 1961 by H. A. Watson of Bell Telephone Laboratories (Bell Telephone Laboratories, 1961; Lee *et al.*, 1985). Since then fault-tree analysis has been used as a safety and reliability tool for complex systems.

The fundamental concept of fault-tree analysis is to construct a structured logic diagram (fault tree) to represent the function of a physical system. The fault tree establishes the relationship between component reliability and system reliability. This logic diagram is constructed using event symbols and logic symbols, often the AND and OR gates.

Fault-tree analysis, which can efficiently focus on the basic events which contribute to a system failure, is an effective tool for reliability and safety analysis, especially as systems become more complex.

A fault-tree analysis may consist of the following four steps (Lee *et al.*, 1985):

1. System definition.
2. Fault-tree construction.
3. Qualitative evaluation.
4. Quantitative evaluation.

The probability of a particular event (such as a system breakdown) can be analyzed with the basic probabilistic information. Component failure rate uncertainty may be modelled using different probability distributions; for

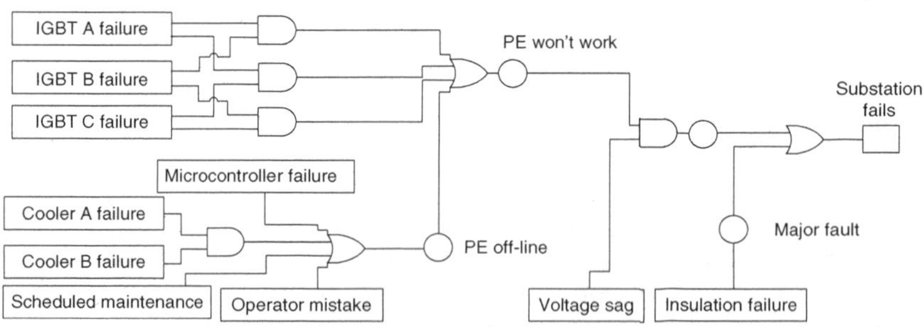

5.19 A fault tree for the solid-state shunt substation sag suppressor, reproduced based on Smith and Atcitty (2009).

example, the Monte Carlo or importance sampling methods can also be used to reduce the computational burden.

Optimization can be performed based on a fault-tree model to decide how to improve reliability and/or reduce costs. Reliability analysis helps to guide reliability improvements in hardware and software technologies, including condition monitoring, prognostics and health management. An example of a fault tree for the reliability evaluation of a solid-state shunt substation sag suppressor is shown in Fig. 5.19 (Smith and Atcitty, 2009).

5.7.4 Condition monitoring and reliability improvements

Condition monitoring of power electronics is useful for diagnosis and for predicting system health. It can also cover all levels from power electronic devices to the whole system. The following parameters may be monitored (Yang *et al.*, 2010):

- On-state voltage of the power semiconductors.
- Waveforms of system voltages and currents.
- Thermal resistance and temperature.
- Switching time of the power semiconductors.
- Gate signals.

Based on system monitoring, diagnostics and prognostics, condition-based maintenance can be implemented to effectively improve the system reliability.

Some important issues affecting the reliability of a power electronic system may include the following:

- Converter load conditions, including worst-case load situations.
- Protection and control under various system fault scenarios.
- Possible interaction among the components in the power circuit and the converter drive and control circuits.

Consequently, the reliability of power electronic systems may also be increased by:

- increasing the quality of the components and the systems;
- decreasing electric and heat loads; and
- appropriate operation, control and protection methods.

Redundancy allows the system to maintain operation under conditions when a system without redundancy would fail.

The effectiveness of redundancy is measured by the reliability increase coefficient:

$$k_p = \frac{P_R}{P_{NR}} \quad \text{or} \quad k_p = \frac{Q_{NR}}{Q_R} \approx \frac{\lambda_{NR}}{\lambda_R} \qquad [5.3]$$

where P_R, P_{NR} are reliability; Q_R, Q_{NR} are unreliability, λ_R, λ_{NR} are failure rate of the redundant system and the non-redundant system, respectively.

5.8 Conclusion and future trends

Offshore renewable energy technologies are developing rapidly. Most of the next-generation large wind turbines are expected to be applied in large wind farms, possibly situated offshore. Higher-power wind turbines in the range up to 10 MW will be installed in the near future. Future large renewable energy-based power generation systems may need not only to have the capability of supporting the grid voltage, VAR control and fault ride-through, but also to provide power dispatch and regulation for the grid. They would become active power plants, instead of simple energy sources. Renewable power plants will be required to perform some duties of conventional power plants, such as grid ancillary services. Power electronic systems will play a very important role in the realization of these capabilities.

The development of power electronic technologies will enhance the role of offshore renewable power generation systems to provide better controllability, power quality, and increase overall system efficiency. Medium-voltage electrical systems may be used for optimizing multi-MW power conversion systems to reduce the cost, weight and volume of the system as well as to reduce system losses. Multi-converter topology provides good fault tolerance, good controllability and good fault-ride-through capability.

The following possible directions for further research in power electronics could be important for enhancing the performance of the direct drive wind and marine power conversion/transmission system.

- New semiconductor devices with low loss, high efficiency and power density.
- Power electronic system topologies for medium-voltage and power transmission applications.
- Reliability improvement of power electronic systems.
- System monitoring, fault diagnosis and tolerance operation.

5.9 References

Anzawa, T., Yu, Q., Shibutani, T. and Shiratori, M. (2007), 'Reliability evaluation for power electronics device using electrical thermal and mechanical analysis', *9th Electronics Packaging Technology Conference, 2007*. EPTC 2007, pp. 94–9.

Balachandran, S. and Lee, F.C.Y. (1981), 'Algorithms for power converter design optimization', *IEEE Transactions on Aerospace and Electronic Systems*, Vol. **AES-17**, No. 3, pp. 422–32.

Bell Telephone Laboratories. (1961), *Launch control safety study, Section VII*, vol 1, Bell Telephone Labs., Murray Hill, NJ, USA.

Blaabjerg, F., Iov, F., Chen, Z. and Ma, K. (2010), 'Power electronics and controls for wind turbine systems', *Proceedings of the IEEE Energy Conference and Exhibition*, ENERGYCON 2010. IEEE Press, pp. 1–12.

Busquets-monge, S., Soremekun, G., Hefiz, E., Crebier, C., Ragon, S., Boroyevich, D., Gurdal, Z., Arpilliere, M. and Lindner, D.K. (2004), 'Power converter design optimization', *IEEE Industry Applications Magazine*, Vol. **10**, No. 1, pp. 32–8.

Chen, Z. (1997), *Advanced Wind Energy Converters Using Electronic Power Conversion*, Ph.D. thesis, University of Durham, England, 1997.

Chen, Z. (2004), 'Compensation schemes for a SCR converter in variable speed wind power systems', *IEEE Transactions on Power Delivery*, Vol. **19**, No. 2, April 2004, pp. 813–21.

Chen, Z. and Spooner, E. (1995), 'A modular, permanent-magnet generator for variable speed wind turbines', *IEE International Conference EMD'95*, Conference Publication No. 412, pp. 453–7.

Chen, Z. and Spooner, E. (1998), 'Grid interface options for variable-speed, permanent-magnet generators', *IEE Proceedings – Electronic Power Applications*, Vol. **145**, No. 4, July 1998, pp. 273–83.

Chen, Z. and Spooner, E. (2001a), 'Voltage source inverters for high-power, variable-voltage DC power sources', *IEE Proceedings – Generation, Transmission and Distributions*, Vol. **148**, No. 5, September 2001, pp. 439–47.

Chen, Z. and Spooner, E. (2001b), 'Grid power quality with variable-speed wind turbines', *IEEE Transactions on Energy Conversion*, Vol. **16**, No.2, June 2001, pp. 148–54.

Chen, Z. and Spooner, E. (2003), 'Current source thyristor inverter and its active compensation system', *IEE Proceedings – Generation, Transmission and Distributions*, Vol. **150**, No. 4, July 2003, pp. 447–54.

Chen, Z., Spooner, E., Norris, W.T. and Williamson, A.C. (1998), 'Capacitor-assisted excitation of permanent-magnet generators', *IEE Proceedings – Electronic Power Applications*, Vol. **145**, No. 6, November 1998, pp. 497–508.

Dai, J., Xu, D. and Wu, B. (2009), 'A novel control scheme for current-source-converter-based PMSG wind energy conversion systems', *IEEE Transactions on Power Electronics*, Vol. **24**, No. 4, April 2009, pp. 963–972.

Deng, F. and Chen, Z. (2009), 'Power control of permanent magnet generator based variable speed wind turbines', 12th International Conference on Electrical Machines and Systems (ICEMS 2009), Tokyo, Japan, 15–8, November, 2009.

Eichler, M., Maibach, P. and Faulstich, A. (2008), 'Full Size Voltage Converters for 5 MW Offshore Wind Power Generators', EWEC2008 proceedings.

Hemmelmann, J. (2008), 'UpWind Transmission & Conversion', EWEC 2008 Session BW2 '"UpWind', http://www.upwind.eu/pdf/Jan%20Hemmelmann%20.pdf.

http://www.landartgenerator.org/readwater6.html

http://www.marineturbines.com/3/news/article/9/seagen_completed__world_s_first_megawatt_scale_tidal_turbine_installed/

http://www.reuk.co.uk/La-Rance-Tidal-Power-Plant.htm

Hu, W., Chen, Z., Wang, Y. and Wang, Z. (2009a), 'Low voltage ride-through of variable speed wind turbines with permanent magnet synchronous generator', Proceedings of International Conference on Ecological Vehicles and Renewable Energy Sources, EVER2009. 2009.

Hu, W., Chen, Z., Wang, Y. and Wang, Z. (2009b), 'Flicker mitigation by active power control of variable speed wind turbines with full-scale back-to-back power converters', *IEEE Transactions on Energy Conversion*, Vol. **24**, No. 3, September 2009, pp. 640–9.

Inoue, S. and Akagi, H. (2007), 'A bidirectional isolated DC–DC converter as a core circuit of the next-generation medium-voltage power conversion system', *IEEE Transactions on Power Electronics*, Vol. **22**, No. 2, pp. 535–42.

Iov, F., Blaabjerg, F., Clare, J., Wheeler, O., Rufer, A. and Hyde, A. (2010), 'UNIFLEX-PM-a key-enabling technology for future European electricity networks', *EPE Journal*, Vol. **19**, No. 4, pp. 6–16.

Kamas, L.A. and Sanders, S.R. (1996), 'Power electronic circuit reliability analysis incorporating parallel simulations', *IEEE Workshop on Computers in Power Electronics*, pp. 45–51.

Krug, D., Bernet, S., Fazel, S.S., Jalili, K. and Malinowski, M. (2007), 'Comparison of 2.3-kV medium-voltage multilevel converters for industrial medium-voltage drives', *IEEE Transactions on Industrial Electronics*, Vol. **54**, No. 6, pp. 2979–92.

Lee, W.S., Grosh, D.L., Tillman, F A. and Lie, C.H. (1985), 'Fault tree analysis, methods, and applications – a review', *IEEE Transactions on Reliability*, Vol. R-**34**, No. 3, pp. 194–203.

Leijon, M., Hans Bernhoff, H., Ågren, O., Isberg, J., Sundberg, J., Berg, M., Karlsson, K.E. and Wolfbrandt, A. (2005), 'Multiphysics simulation of wave energy to electric energy conversion by permanent magnet linear generator', *IEEE Transactions on Energy Conversion*, Vol. **20**, No. 1, March 2005, pp. 219–24.

Li J., Wu, X., Hu, S., Zhao, B. and Xu, H. (2009), 'Direct-drive wind power generator system based interleaved boost converter', *Proceedings of ISES World Congress 2007* (Vol. I–Vol. V), Vol. **6**, pp. 2301–4.

Michalke, G., Hansen, A.D. and Hartkopf, T. (2007), 'Control strategy of a variable speed wind turbine with multipole permanent magnet synchronous generator', EWEC2007 Proceedings.

Mueller, M.A. (2002), 'Electrical generators for direct-drive wave energy converters', *IEE Proceedings – Generation, Transmission and Distribution*, Vol. **149**, pp. 446–56.

Pickert, V., Zahawi, B. and Kalpoktsoglou, D. (2011), 'Forced commutation controlled series capacitor (FCSC) circuit applied to stand-alone wave energy conversion buoys', *Journal of Marine Engineering and Technology*, Vol. **10,** No. 1, pp. 15–23.

Polinder, H., Damen, M.E.C. and Gardner, F. (2004), 'Linear PM generator system for wave energy conversion in the AWS,' *IEEE Transactions on Energy Conversion*, Vol. **19**, No. 3, September 2004, pp. 583–589.

Senturk, O.S., Helle, L., Munk-Nielsen, S., Rodriguez, P. and Teodorescu, R. (2009), 'Medium voltage three-level converters for the grid connection of a multi-MW wind turbine', *Proceedings of EPE*, pp. 1–8.

Smith, M.A. and Atcitty, S. (2009), 'Power Electronics Reliability Analysis', Sandia Report, SAND2009–8377, http://prod.sandia.gov/techlib/access-control.cgi/2009/098377.pdf, accessed on 27 December 2010.

Tavner, P. (2012), *'Reliability in Renewable Energy Systems'*, presentation at the third international symposium on Power Electronics for Distributed Generation Systems, Aalborg, Denmark.

Tenca, P., Rockhill, A.A. and Lipo, T.A. (2007), 'Wind turbine current-source converter providing reactive power control and reduced harmonics', *IEEE Transactions on Industry Applications*, Vol. **43**, No. 4, July/August 2007, pp. 1050–1060.

Troedson, A. and Desai, P. (2010), 'A comparison of low voltage and medium voltage wind turbine drive trains', EWEC2010 Proceedings. EWEC 2010, 22 April, Warsaw.

Yang, S., Xiang, D., Bryant, A., Mawby, P., Ran, L. and Tavner, P. (2010), 'Condition monitoring for device reliability in power electronic converters – a review', *IEEE Transactions on Power Electronics*, Vol. **25**, No. 11, pp. 2734–52.

Zeng, X., Chen, Z. and Blaabjerg, F. (2007) 'Efficiency comparisons of full-size converters for large variable-speed wind turbine systems', the 13th European Conference on Power Electronics and Applications, EPE2007, September 2007.

Zhang, D., Wang, F., Burgos, R., Lai, R., Thacker, T. and Boroyevich, D. (2008), 'Interleaving impact on harmonic current in DC and AC passive components of paralleled three-phase voltage-source converters', *Twenty-Third Annual IEEE Applied Power Electronics Conference and Exposition*, APEC 2008. 24–28 February 2008, pp. 219–25.

Part II

Applications: wind and marine

6
Wind turbine drive systems: a commercial overview

E. DE VRIES, Rotation Consultancy, The Netherlands

DOI: 10.1533/9780857097491.2.139

Abstract: Over three decades modern wind power development is characterised by huge advancements in turbine scaling, drive technology, switching rotor fixed-speed to variable-speed operation, advanced load-control and grid integration. Larger rotors per megawatt aims at much lower cost of energy (CoE) and achieving accelerated cost parity with conventional power plants. Hybrid drive from the late 1990s expanded the choice between high-speed geared and direct drive. Double fed induction generators (DFIGs) were long been predicted to lose out from permanent magnet generators (PMG), while by mid-2010 PMG-based direct drive was predicted to become the ultimate drive solution. DFIG proves more enduring than predicted, and uncertainty on future rare earths price development and fluctuations has turned medium-speed requiring smaller PMGs into a main trend. Hydraulic-drive solutions offer variable rotor speed with fixed generator speed and appear on the increase.

Key words: CoE, variable-speed, DFIG, PMG, medium-speed, hydraulic-drive.

6.1 Introduction

This chapter focuses on commercial wind turbine drive systems, reviewing both common and uncommon designs, and solutions developed both early in the history of modern wind power and nearer the present time. The ultimate wind turbine drive system solution has not yet been developed, nor is it expected to emerge in the foreseeable future. In recent years a varying and unpredictable combination of technical, geopolitical and other key factors have impacted wind technology and these factors have been subject to continuous change. This chapter therefore attempts to point out key developments that have already occurred, current trends and what the combination of history and the present may imply for the future of wind power.

6.1.1 Types of wind turbine

The primary function of a traditional windmill is to convert power in the wind into mechanical power, which is directly employed for specific 'end use' applications like pumping water, milling grain or sawing wood. A key characteristic of windmills is thus that power generation is typically combined with direct use of the energy in the same confined area. Some early designs like the Persian Windmill date back to at least 600–900 AD. Modern wind turbine drive systems, by contrast, as a primary function convert power in the wind into mechanical power, and as a next step into electric power fed into the electricity network.

Apart from some fundamental functional differences between windmills and wind turbines, for each category a similar main subdivision can be made into horizontal-axis and vertical-axis wind turbines. The word 'axis' refers to the main drive shaft and thus rotor orientation. Horizontal-axis turbines clearly dominate the wind market, but are also inherently more complex as they require a yaw system to continuously direct the rotor towards the prevailing wind direction. Vertical-axis wind turbines are by comparison rather uncomplicated and as a key system and operating advantage can draw wind from all directions. However, a main disadvantage is their much lower maximum aerodynamic efficiency that lies in the 35–38% range as compared to 45–50% or higher for horizontal-axis equivalents. Finally, regarding power output control strategy, both categories can either be fitted with blades mounted at a fixed blade angle or employ pitchable equivalents capable of rotating around a blade longitude axis.

6.2 Early geared wind turbine drive systems

'Danish Concept' turbines are often associated with the birth of modern wind industry in the late 1970's. Key to the remarkable success story was a practical bottom-up and 'learning by doing' approach, which enabled several pioneering Danish firms to develop into market leaders and many years of world wind market dominance for Denmark.

6.2.1 'Danish Concept' wind turbines with fixed-speed induction generators

In the late 1970s Danish wind pioneers introduced a relatively uncomplicated wind turbine design concept that became highly successful and is known as the 'Danish Concept'. Its main characteristics include a non-integrated high-speed geared drive system, a fixed-speed directly grid-connected induction or asynchronous generator, and fixed-angle rotor blades with movable tip brakes. The common four-pole generator together with 50 or 60 Hz grid frequency determines a fixed generator speed, while the corresponding rotor speed is determined by a fixed gearbox ratio. If, for example, the generator rated speed in a 50 Hz grid network is 1500 rpm, and the gearbox step-up ratio i = 1:75, the corresponding rotor speed is 20 rpm. The technology evolved progressively (bottom-up product and technology development): the initially small 20–35 kW-size 'Danish Concept' stall-regulated wind turbines (known as 'Classic-stall' systems) gradually grew in size until a practical 1.5 MW maximum power rating was reached in the late 1990s. However, several factors combined to mark the end of this once leading wind technology based upon a passive output limitation principle.

First, the so-called stall point where power output is limited depends upon specific environmental conditions, with 'power overshoot' being a well-known and infamous operational phenomenon. Power overshoot is defined as a maximum output excursion in excess of the rated power rating. Multiple wind industry sources have reported overshoot of up to 20% during combined cold-temperature high-wind conditions, but this unwelcome phenomenon has particularly affected specific makes and models of turbine. Regular wind turbine power overshoot can seriously reduce the life of key drive components such as the gearbox and generator; to counter this the operators often had to reset the blade angles twice a year, in the early summer and again in the autumn.

A second inherent constraint directly linked to the nature of the passive fixed-speed Classic-stall concept is that it excludes the advanced power electronic control capabilities that are now common in modern turbine control systems. Examples of such capabilities include active grid support during grid failure (automated) power factor adjustment, and the option of active output reduction during night hours to curb aerodynamic noise.

6.2.2 Developments in fixed-speed turbine models

Pole-switchable generators and active-stall control

Some fixed-speed turbine models are fitted with a pole-switchable 4/6-pole induction generator. This enables (six-pole mode) turbine operation at reduced rotor speed in the lower wind speed ranges up to about 7 m/s, a measure specifically aimed at aerodynamic noise limitation. Several later developments, especially large fixed-speed turbine models, feature a more advanced control system known as Active-Stall Control. The pitchable rotor blades enable a near constant rated output level from rated speed to stopping speed (cut-out). A second key advantage compared to Classic-stall is that Active-Stall Control eliminates the need for bi-annual blade angle adjustment. One of the main reasons that prompted suppliers to switch from Active-Stall to pitch-controlled variable speed is improved noise characteristics at high-wind speeds. And, for larger turbines, the accompanying reduced blade tip deflection is also regarded a major benefit, along with a potential reduction of dynamic loads.[1]

Pitch-controlled fixed speed

With the combination of fixed speed and stall power limitation being semi-standard in wind technology of the 1980s and 1990s, an alternative pitch-controlled fixed-speed solution also enjoyed considerable popularity. Vestas is perhaps best known for a range of such turbines up to a largest 500 kW V39500kW model (note that the type model is V30-500 kW). However, an often quoted disadvantage is that gusty wind conditions can cause high gearbox and drive system impact loads. The key reason for this is that the pitch system is only engaged after a gust of wind has already hit the rotor and the resulting impact load has passed on directly into the drive gearbox and generator.

Rotor current control

During the first half of the 1990s Weier of Germany developed a semi-fixed-speed system comprising an induction generator fitted with fast-reacting power electronics. Called Rotor Current Control or RCC, the system allows an instant temporary 10% rotor speed increase when a gust hits the rotor. 'Extra' energy contained in a gust is temporarily stored in rotor speed acceleration, and released again once the blade angle has been readjusted and the rotor returns to its rated speed level. Under the trade name OptiSlip®, Vestas further developed a functionally comparable system that was introduced in a number of turbine types from 600 to 1.75 MW. Operational experiences, especially in terms of dynamic load reduction and improved power quality, were reported as quite positive. However, the OptiSlip generator is directly grid-connected in analogy with fixed-speed stall turbines, which excludes advanced control options that are a standard feature of modern variable-speed wind turbines.

At the moment Suzlon of India is (one of) the last remaining major supplier(s) to still apply such electronically adjustable generator slip technology in a S.8.8-21 MW mode. Named Flexi-Slip, the technology provides a slip percentage up to 16.67%.

6.2.3 Early variable-speed concepts

Variable-speed wind turbine application dates back to the 1920s and 1930s when thousands of small wind turbines powered rural homes and farms in the US and elsewhere. Typical installations were in the 0.5–10 kW range, and the power produced was stored in batteries.

During the early 1980s the first commercial kW-size geared variable-speed wind turbine concepts were introduced by a selected number of wind pioneers including the Dutch company Lagerwey and Enercon of Germany.

Much larger variable-speed research prototypes like the 1 MW two-blade Stork-FDO NEWECS-45 and Boeing's 3.2 MW MOD-5 also originate from the 1980s.

During the early 1990s only few variable-speed geared wind turbine models like the two-blade Lagerwey 80 kW LW80 and 250 kW LW30, the 80 kW Enercon E-17/E-18 and 300 kW Enercon E-32 models were commercially available. In 1993 former US company Kenetech introduced the 33M-VS turbine model, hailed at the time as an absolute breakthrough turbine development in terms of efficiency, including wind power production capability of 0.05 US$/kWh.[2] Unfortunately, this ill-fated turbine concept contributed much instead to the company's bankruptcy within a few years.

6.3 Direct drive generators

Direct-driven generators characterised by large diameters and (multi-) megawatt-scale power ratings have been applied for many decades in hydro-electric plants, including high-ranking applications like the Niagara Falls at the US–Canadian border.

In 1992 Enercon introduced the world's first commercial direct drive turbine, the 500 kW E-40 with rotor diameter of 40.3 m. In contrast with hydro-electric applications where load variations are relatively minor, wind turbines are exposed to continuously varying loads due to the stochastic character of the wind itself. Wind turbine operation is also typically characterised by many hours of partial load, large and continuous power output fluctuations, and continuously varying thermal electrical and mechanical generator loads and loading conditions.

6.3.1 External field excitation

Regarded as a technical sensation over two decades ago, the E-40 featured a multi-pole ring generator, developed in-house, with external field excitation and an outer diameter of around 4 m. This E-40 turbine model achieved widespread recognition for Enercon founder and owner Aloys Wobben, especially because he had succeeded in adapting and optimizing ring generators for wind turbine use. Dutch wind expert Henk Hutting commented, for instance, that with the E-40 generator concept Enercon had struck the right balance between costs and efficiency, with a design focus on balancing copper use and necessary manufacturing tolerances. Hutting also cited the main outcomes of an earlier direct drive generator study conducted jointly by former Dutch wind turbine supplier WindMaster and Dutch electric power engineering equipment supplier Holec: 'One main conclusion was that reapplying hydropower generator technology solutions would have resulted into a very high machine cost price, with total copper requirement being one of the key generator cost drivers. A higher copper percentage is thereby a measure of electric magnet field strength'. He said, in conclusion, that Enercon had, for manufacturing reasons, chosen a rather large air gap, which had negatively affected generator efficiency.[3]

Technology developments at Enercon

Enercon achieved a second direct drive milestone in 1995 with the 1.5 MW E-66 prototype, featuring a ring generator with outer diameter of about 5.3 m. This turbine model itself was one of four different 1.5 MW turbine developments supported by the European Commission, which also included a Vestas V63–1.5 MW, Nordtank NTK1500/60, and Tacke TW1.5. This initial E-66 turbine model evolved in the 1.8/2 MW E-66 (1999), 2.0/2.3 MW E-70 (2003) and the 2.0/2.3/3 MW E-82 volume series (2005). The latest addition is the 2.3 MW E-92 (2012). These models are all semi-standard, fitted with an electrically excited air-cooled generator featuring the same outer diameter of about 5.3 m. The exception are strong-wind 3 MW E-82 E3 and E-82 E4 model versions for respectively IEC class IIA and IA, both featuring a new water-cooled generator. The diameter of all in-house manufactured Enercon generators has remained unchanged, but generator axial stack length was increased with the E-82.

Rotor diameters have been increasing with new models since 1995, but in addition, each time rotor diameter is increased a reduction in rotor revolutions becomes necessary to curb aerodynamic noise. An important issue here is the relation power output (P) as a function of torque (T) and rotor speed (n), whereby power is a function of torque multiplied by rotational speed in radians per second. If, thus, for a given power rating rotor speed goes down, the corresponding generator torque has to increase by a similar percentage. A higher torque eventually means more heat production, which places additional demands on the generator heat dissipation capacity.

In 2002 Enercon installed a prototype of the 4.5 MW E-112. It features a huge air-cooled ring generator with an outer diameter of 12 m, comprising a rotor split into two 180-degree sections and a stator electrically subdivided into four 90-degree segments. The result is that each generator stator section functions electrically as if it were an individual generator.

In 2005 the E-112 power rating was upscaled to 6 MW, while in 2007 a second-generation E-126 successor turbine with a largely identical 6 MW generator was introduced. The rotor diameter increased from 114 to 127 m and in 2011 the turbine was upscaled to 7.5 MW, in the usual evolutionary small-step scaling and optimizing company approach. For a new 3 MW liquid-cooled E-101 turbine platform generator length has again been increased but by maintaining a 5.3 m outer diameter. A very latest E-101 sister model is the 2.5 MW E-115 again with water-cooled generator and an enlarged 115-m rotor diameter.

Competing designs

Besides Enercon, at least three other suppliers are known to have developed direct drive turbine generators with external field excitation. US-based Northern Power Systems has for years marketed a 100 kW model, 40 of which have been installed.[4] In 2008 the company switched to a new permanent magnet generator (PMG), developed in-house. In the mid-1990s the former Dutch company Lagerwey introduced a 750kW turbine with an air-cooled generator about 5.5 m in diameter, of which 200–225 have been built. Following Lagerwey's 2003 bankruptcy a new company, Emergya Wind Technologies (EWT), bought the rights and further developed the turbine models into an upgraded 900 kW model version with enlarged rotor diameter but based upon similar technology principles. More recently EWT switched to a new generator design for this model. MTorres of Spain installed a 1.5 MW TWT 1.5 direct drive prototype in 2001, which was later upscaled to a 1.65 MW TWT 1.5.[5] Set against Enercon's cumulative design base since 1992, with a total of over 20 000 direct drive turbines installed (status end November 2012), cumulative competitor numbers are by comparison minor.

6.4 Doubly fed induction generators (DFIGs)

Doubly fed induction generators (DFIGs) were considered a rather obsolete generating technology in the first half of the 1990s. Austrian wind pioneer and Windtec founder Gerald Hehenberger claims that he started developing a DFIG system for variable-speed wind turbines as early as 1993. He also reportedly informed several wind turbine suppliers about the solution from 1994 onwards, but experienced little real interest.[6]

In April 1996, the former Tacke Windtechnik of Germany (now GE) installed a 1.5 MW pioneering pitch-controlled variable-speed prototype called TW 1.5 fitted with a DFIG. The former company DeWind of Germany followed a few months later with a 500 kW prototype.

A key DFIG benefit is that the stator is directly connected to the grid network. Most of the time over 75% of the power generated is directly fed into the grid. The power percentage fed through the converter is proportional to the slip; at synchronous speed, the only power that flows through the converter is what is needed for ensuring the flow of DC excitation current through the rotor windings. DFIG-based electric power conversion systems therefore require only a rather small and thus by comparison inexpensive so-called partial power converter. Fifteen years ago power electronics were, according to wind industry sources, an estimated factor of 8–10 more expensive than today. Tacke's new power conversion solution initially claimed to offer an investment saving of the order of € 75 000 compared to equivalents fitted with a synchronous generator and full-power converter.

During the initial years of DFIG operation there were a number of well-documented generator issues. One was the limited life of the generator rotor brushes, 2–4 months. A second problem was premature bearings damage caused by stray or transient currents, and apparently to a lesser extent also due to wear particles from carbon brushes entering particularly the rear bearing. These issues have now been addressed. Brush life has now been extended to at least 6 months to a year, depending upon operating conditions, turning it into a 'consumable' for exchange during regular service intervals. Perhaps most remarkable of all is that only a few years after their introduction DFIGs had already become a very popular semi-standard in the wind industry.

6.4.1 Impact of new grid requirements on DFIGs

In 2003 German electricity distribution giant E.on Netz GmbH announced new compulsory grid requirements for wind operators in its supply area, which came into force from January 2004. One of the new requirements was that in the event of a major grid problem, wind turbines have to remain grid-connected or 'ride-through', with a built-in capacity for active grid support. In other words, from a grid functioning point of view, wind turbines have to behave like conventional fossil or nuclear power plants. Prior to the introduction date there were considerable fears among wind industry experts that DFIGs would not be capable of meeting E.on's new grid rules. One of the conclusions was that standard DFIGs with a partial or partly rated power electronic converter feeding the rotor coils cannot meet these advanced grid fault ride-through requirements. This is because during grid faults the resulting fault currents would be so large that the power electronics would fail. Additional hardware and controls were needed in order for DFIGs to comply with the grid fault ride-through requirements.[7]

6.4.2 New developments

There are at least two wind industry initiatives aimed at further developing and optimizing DFIGs by eliminating some of the current disadvantages.

Spain's Indar/Ingeteam has developed an innovative solution called Ingecon® Clean Power Series®, a system comprising a DFIG with integrated so-called exciter PMG. This xDFM® is in essence a permanent magnet (PMG) type synchronous machine mechanically coupled to the DFIG inside a common standard machine housing. Like a 'standard' DFIG system, the Ingecon® Clean Power Series® System unit comprises a partial power converter with 25–30% generator rated power. The exciter machine rated capacity is about 16–18% of generator power rating. Its inclusion made it possible to connect the power converter to the rotor of the DFIG, and to the stator of the PMG, thus isolating it from the grid with the stator being the only grid-connected output. To make speed variation possible in a conventional DFIG, the rotor power goes via the power electronic converter to the grid, with the stator directly grid-connected (as is the case with xDFM).

In the Ingecon solution, rotor power goes via the power electronic converter to a permanent magnet machine mounted on the same shaft as the DFIG and creates an additional torque. In this way, all power going into the grid comes via the stator. The xDFM® technology, according to its inventors, offers outstanding grid quality as well as exceptional low-voltage ride-through capabilities. That in turn is claimed to limit torque peaks and reduce mechanical peak loads that would otherwise typically happen during a grid fault or in emergency situations.[8]

A UK group originating in universities and industry has spend about 10 years developing another novel design called brushless DFIG. After several development and testing stages, the group says it has just completed successful testing of a 250 kW machine. One of the key experts involved is power engineering expert Peter Tavner, emeritus professor at the School of Engineering and Computing Sciences at Durham University in the UK.[9]

Regarding DFIG's future prospects, REpower chief technology officer Matthias Schubert said in a 2010 interview: 'In combination with a partial converter (30–35% of the turbine's power rating), this generator type offers the most cost-effective electrical power conversion solution. The solution can also meet any future grid requirement, as in our view, everything needed is already included'.[10] Other experts have expressed different viewpoints reflecting alternative technology preferences and solutions.

Today the largest DFIGs can be found in a variety of wind turbine makes and models, with the largest unit being applied into a 6.15 MW REpower 6 M offshore turbine model. For the last few years the dominant role of DFIGs has been eroded somewhat due to the emergence and growing role of PMGs. Remarkably US giant GE in an opposite development trend announced at the Husum WindEnergy 2012 international wind fair in Germany that it switches from PMG to DFIG for its 2.5–2.75 MW turbine platform. The move follows an opposite switch to PMG 10 years ago and is said to offer both investment cost and total systems efficiency (generator + converter) benefits.[11]

6.5 Low- and medium-speed (MS) geared hybrid concept

Low- and medium-speed (MS) drive system concepts can be described as an intermediate design solution between a conventional fast-speed geared drive – multi-stage gearbox with fast running generator – and direct drive (no gearbox).

Today a limited but growing number of commercial designs are commercially available in the 1–5 MW class, with either a single-stage or two-stage gearbox. The first solution enables a rotor speed step-up ratio of about 1:10 with a special gearbox design approach sometimes referred to as '1.5-stage', whereas two-stage gearboxes offer a maximum step-up ratio of approximately 1:40. A faster rotating two-stage medium-speed drive system typically increases gearbox weight and complexity compared to a single-stage system, but the generator is usually somewhat lighter and cheaper.

6.5.1 Multibrid technology

The first medium-speed product development by the German engineering consultancy Aerodyn Energiesysteme originates from the mid-1990s. Advertised as a 'quantum leap' into the multi-megawatt class, their 5 MW Multibrid offshore wind turbine was first presented at the international Hannover industrial fair in April 1998. The name Multibrid stands for Multi- and -hybrid, and the technology was based on the outcome of 'extensive comparable studies of various drive train designs', focused at critical components and key design variables. In one study the researchers compared three standardized 1.5 MW variable-speed units, each with a 65 m rotor diameter with a main focus at design features and operating economics. Top head mass (THM, nacelle + rotor) and highest possible systems reliability were assessed as the two single most important decision criteria for the new drive train concept.

In its current form the 5 MW Multibrid M5000 comprises a single main bearing with a single-stage planetary-type gearbox and a medium-speed PMG with a nominal speed of 148 RPM. All three main components are integrated into a compact and stiff single load-transmitting structure.

The Multibrid concept includes extensive systems integration. The main slow-running bearing, gearbox and generator do not, for instance, have their own separate housings. They are instead fitted directly into the main cast iron machine structure, with even the generator bearings being part of the gearbox assembly. According to the designers, compactness, combined with the modest dimensions of the 28-pole synchronous 3 kV PMG, enables full enclosure of sensitive components like generator windings against the aggressive offshore environment.[12]

Multibrid technology has been licensed to a company called Multibrid (now AREVA Wind) and to WinWinD of Finland. The first 5 MW Multibrid prototype was installed during 2004 by Multibrid (now AREVA Wind), followed by three additional onshore prototypes. Another 6 units have been installed in Germany's first offshore wind farm Alpha Ventus by AREVA Wind. In total 120 units are planned in the North Sea until the end of 2013. A new 5 MW M5000 version with enlarged 135 m rotor has also been announced.

WinWinD is licensed to develop Multibrid-type wind turbines up to 3 MW power rating and already markets two different medium-speed models of 1 and 3 MW. Like the larger M5000, both turbine models incorporate a fully integrated cast main structure and a PMG.

In 2011 WinWind presented a totally new 3 MW WinWind 3 design offering 50% nacelle mass reduction, and this time incorporating a separate two-stage gearbox plus, again, a PMG. Opinions differ as to whether this turbine should still be called a 'true Multibrid'.

6.5.2 Gamesa Eolica's medium-speed wind turbines

In late 2008 Gamesa Eolica of Spain erected a medium-speed 4.5 MW wind turbine G128–4.5 MW prototype at an R&D wind farm. Key product development goals were maximum drive system reliability and the ability to employ transportation and erection equipment logistics similar to the 2 MW Gamesa G8X-2.0 MW series. Gamesa's reliability goal resulted in a semi-integrated medium-speed drive system. The Compact Drive drive system comprises a main shaft with two main bearings, a two-stage planetary-type gearbox with 1 ÷ 37.9 speed-up ratio, and PMG. The generator is directly attached to the gearbox with a flanged connection, whereas the complete drive system is split into separate main components for ease of component exchange.

Besides striving for maximum reliability, additional requirements were optimized systems flexibility and stability, together aimed at 'setting new standards with regard to Life Cycle Costs (LCC) performance'. According to Gamesa, major direct drive system limitations include complex and costly transport and erection logistics due to generator size and weight. These could be avoided by choosing a medium-speed system. Other Gamesa medium-speed drive system preferences include overcoming size and weight issues linked to direct drive, and eliminating the 'trouble-prone' high-speed gear stage of three-stage gearboxes. It was important for the designers not to compromise overall systems flexibility by choosing to eliminate the high-speed gear stage in order to improve reliability.[13] Gamesa has also announced, based upon the initial G128–4.5 MW medium-speed turbine, a 4.5 MW version with enlarged rotor and two new offshore-dedicated 5 and 7 MW turbines.

6.5.3 Additional product development trends

On 30 March 2011 Vestas announced a new 7 MW V164-7.0 MW offshore wind turbine fitted with a three-stage planetary gearbox that is connected to a generator with a rated speed of 400 RPM.[14] However, in an 2 October

2012 press release, the company announced that the final product had been upgraded to an 8 MW V164–8.0 MW, stating:

> From the very beginning the V164 has been developed with a potential of increasing the turbine size. As we progressed in the technology development it was clear that an 8 MW version of the turbine will offer lower cost of energy and at the same time keep the reliability and structural integrity of the turbine unchanged.

South Korean shipbuilder DSME is another party that also presented a new 7 MW three-bladed medium-speed offshore turbine in development in 2011. Drive system specifications along similar lines to the V164–7.0 MW include a main shaft with two bearings, a two-stage gearbox and medium-speed 3.3 kV PMG. The turbine will be fitted with a 160 m rotor. Regarding drive system preference a DSME expert explained: 'Prior to the design phase start DSME conducted a drive system comparison. The main outcome was that energy generating costs for direct drive and medium-speed geared drive systems are roughly similar. We chose a main shaft with two bearings solution as this limits the risk of rotor-induced loads being passed on to the gearbox'.[15]

Spanish wind industry generator and power converter supplier Ingeteam/Indar Electric has observed two different tendencies for low- and medium-speed systems in turbines ranging from 2.5 to 5 MW; a single-stage gearbox and about 150 nominal generator rpm, and a faster-running double-stage gearbox with about 450 rpm for the generator. In their view a combination with a PMG, rather than another type of generator, might offer an optimal overall choice for both direct drive and medium-speed drive systems, largely due to (more compact) dimensioning considerations.[16] At the moment there appears to be a strong market preference for two-stage gearbox systems purely for lifecycle energy costs reasons, because the faster-running generator has a much reduced requirement for expensive rare earth elements (REEs). In practice there is also a trade-off between reliability, complexity and often underestimated specific company preferences.

6.6 Permanent magnet generators (PMGs) in direct drive wind turbines

PMGs have been integrated in large direct drive turbines since at least 1997, whereas commercial production (on a limited scale) commenced around 2003. In 1990 wind pioneer Professor Friedrich Klinger founded Forschungsgruppe Windenergie (wind turbine development group) at the Saarbrücken University of Applied Sciences in Germany. This early initiative has resulted in a number of innovative turbine developments, some with major impact on the global wind industry. In 1995 the group started, among others, on the design of a 600 kW direct drive turbine incorporating an innovative passive air-cooled PMG. The Genesys prototype was installed in 1997. In 2000 Vensys Energiesysteme was founded, and 3 years later a 1.2 MW Vensys 62 prototype was installed. This model was upgraded to 1.5 MW, with initial licensee GoldWind of China now a majority shareholder. GoldWind has become the world's largest supplier of PMG-type direct drive turbines, with a 2011 track record of at least 3000 operational 1.5 MW units.[17,18]

Dutch direct drive pioneer Lagerwey commenced development of a 1.5 MW/2 MW PMG-type direct drive turbine model named Zephyros in 2000. The later independent supplier Zephyros filed for bankruptcy in 2005, but the innovative Zephyros technology has been disseminated to at least one Chinese company, XEMC, and to South Korean company STX.

Currently a number of different suppliers are developing and some are already marketing commercially PMG-type direct drive turbines with power ratings from 1.5 to 3 MW and soon 6 MW. The first multi-megawatt class fast-speed geared wind turbine fitted with PMG was a 2.5 MW GE 'technology demo' prototype turbine in 2006. Several other suppliers – including Vestas, Unison and PowerWind – have also developed fast-speed geared turbine models fitted with PMGs. Furthermore, a growing number of turbine models with PMG at various project completion stages in the 5–8 MW class are being developed.

Leading suppliers that either have developed and operate prototypes already or have announced PMG-type 6 MW direct drive offshore turbines include Alstom, GoldWind, and Siemens Wind Power (SWP). Nordex was already at an advanced stage of development with a 6 MW prototype, but in April 2012 announced the cessation of new offshore business. GE has developed a 4.1 MW direct drive turbine. In March 2011 GE acquired leading generator and power converter developer and manufacturer Converteam, which was well known as a leading PMG developer/supplier. US-based Northern Power Systems is developing an 8 MW direct drive turbine with 175 m rotor diameter. An interesting feature is the choice of a segmented stator design, a feature that aims at both industrialisation of production and enhanced system redundancy capability.[19] Finally, Sway AS of Norway is developing an unusual 10 MW direct drive turbine with a fully exposed spoke-type axial-flux generator, featuring a diameter of 25 m.[20]

PMGs currently enjoy a likely 100% market share in the low- and medium-speed segment. PMG dominance in large low- and medium-speed wind turbines can be explained by their reduced size and mass and perceived superior capability to meet future grid code demands compared to classic generators. The largest medium-speed offshore wind turbine incorporating PMG was the 10 MW Clipper Britannia, until the project was shelved in 2011.

Within about a decade, PMGs have turned into a genuine wind industry trend, with global power engineering firms among the powerful new entrants in this area. The single main bottleneck hampering even faster global reach are the current high prices and even more the price volatility of the REEs used for the magnets and the long-term wind industry uncertainty this continues to create.

6.7 Alternative technologies and power conversion

While DFIG's and PMG dominate the wind turbine generator market, some suppliers continue to prefer either classic synchronous generators or induction/asynchronous generators. Furthermore, range of mechanical, mechanical-hydraulic and full-hydraulic drivetrain solutions by different technology suppliers are in development. These solutions enable a combination of variable rotor speed and fixed generator speed, thereby eliminating the need for a power electronic converter.

6.7.1 Alternatives to DFIGs and PMGs

The majority of today's fast-speed geared wind turbines are either fitted with a DFIG or a synchronous PMG. SWP is an exception among leading wind turbine suppliers as it chose a four-pole squirrel-cage induction generator with full converter instead. These generators are applied in the Siemens onshore/offshore 2.3 MW and offshore 3.6 MW geared volume platform models. During 2004 SWP (formerly Bonus Energy until October 2004) chief technical officer Henrik Stiesdal explained the reasons for the company's choice of a brushless (squirrel-cage type) induction generator, combined with a full converter.

> Induction-type generators are widely used, known to be relatively simple, robust and cost-effective. In fact, since 1998, Bonus has had a 1 MW machine operating with variable-speed pitch control and a full converter. We are of the opinion that a full converter is better suited for offshore applications than the doubly fed systems common today. With an induction generator, we avoid the problem of slip rings, and it allows the best possible grid compliance. In addition, the generator concept is similar to the technology we have already applied for over 20 years in all our fixed-speed designs, from the first 22 kW machine to the latest 2.3 MW turbine.'[21]

Wind Energy Solutions (WES) of the Netherlands is a small supplier that applies induction generators in its 50, 80 and 250 kW turbine models. The technology was acquired from former Dutch supplier Lagerwey, which had built variable-speed wind turbines incorporating induction generators since the early 1980s.

For wind turbines the induction generator application is largely limited to four-pole machines as the electrical efficiency goes down with a higher pole number. Being a semi-industry standard, prices quoted for induction generators are slightly cheaper than for synchronous generator equivalents, but the electrical efficiency is also a bit lower.[22]

Interestingly the price explosion of REEs, used for common dysprosium and neodymium-type PMG magnets, appears in recent years to have initiated a revival of classic electrically exited synchronous generators. A relatively modest increase in generator dimensions and mass is here regarded as a fair price to pay for reduced dependence on REE price volatility with the added uncertainty about future supply. Synchronous generators operating with variable speed require a 'full converter', necessary for converting power generated with continuously changing frequency into grid-compliant power for feeding into the grid network.

Indian supplier Kenersys applies a synchronous generator with external field excitation in its 2 and 2.5 MW wind turbine models. Former managing director Andreas Reuter argued during an expert technical discussion in 2007/8 that the operational track record of PMGs in wind turbines at that time was limited to about five years. Reuter, who is currently a wind energy university professor at the Gottfried Wilhelm Leibniz Universität Hannover in Germany, further said he preferred incorporating classic synchronous generators with electrical field excitation for long-term operational risk-reducing considerations.

6.7.2 Variable rotor speed with fixed-speed generator

In wind turbines operating with variable rotor and generator speed, a power converter is required to convert generator power of continuously varying frequency into grid-compliant 50 Hz/60 Hz grid network power. In the past decade different technology suppliers have either introduced or are developing systems enabling the wind turbine rotor to operate at variable speed and the generator at fixed speed. Such functionally comparable system solutions enable direct grid network connection similar to conventional power plants, and are claimed to offer the highest possible grid power quality. A main shared benefit is the elimination of the power converter that would otherwise be required. Moreover, the common choice of a medium-voltage generator under certain network conditions enables a medium-voltage transformer to be eliminated.

6.7.3 Power conversion systems

Power conversion systems can be subdivided into mechanical-hydraulic (Voith WinDrive & Voith-RENK AeroDrive, and AMSC Windtec superGEAR), mechanical (IQgear and VMT Technologies), electro-mechanical (SET) and fully hydraulic (Mitsubishi). A disadvantage often claimed especially for power conversion systems involving hydraulics is worry about high internal losses resulting in inferior overall efficiency compared to power systems based upon mechanical-electric-electronic power conversion.

The longest and most comprehensive track record has been achieved with Voith of Germany's hydrodynamic WinDrive incorporated into 2 MW DeWind turbines from 2006. Another much larger suggested 6.5 MW WinDrive application is for a new BARD 6.5 offshore turbine, whereby a gearbox with two output shafts is connected to two WinDrives and two matching medium-voltage synchronous generators. Latest product development is the AeroDrive for turbines up to 8 MW.

The superGEAR hydrostatic system is currently offered for AMSC design 2 and 3 MW turbines, but the 2011 track record was limited to two 2 MW prototypes installed in 2008. Like Voith's WinDrive, the superGEAR system comes with a two-stage gearbox and medium-voltage synchronous generator.

IQgear technology developed in Israel involves a fully mechanical system with a continuously variable internal speed ratio. Its present status is unknown.

VMT Technologies has developed a fully mechanical system comprising a belt system with continuously variable ratio in combination with a two-stage gearbox.[23]

Sustainable Energy Technologies (SET) is one of the latest entrants in this section of the wind industry, with an innovative integrated solution called 'electro-mechanical differential system'.

Finally, technology firm Artemis has developed a fully hydraulic power conversion system. In late 2010 industrial giant Mitsubishi Heavy Industries (MHI) of Japan acquired the UK-based company. The following year MHI announced a 7 MW offshore-dedicated SeaAngel wind turbine with full hydraulic transmission based upon Digital Displacement® technology, and twin fixed-speed electrically excited synchronous generators.[24]

6.8 Reliability, availability and total systems efficiency

There are many and often rather different views on wind turbine reliability, availability and total systems efficiency. Geared wind turbines suffer an unfortunate reputation for unreliability compared to direct drive due to an envisaged trouble-prone gearbox. Many further believe that simply removing the combination of gearbox and fast-speed generator, and exchanging this for a direct drive ring generator, would automatically create a highly reliable wind system. It is a fact that there have been in the past many issues with premature wind turbine gearbox failures. It is also true that considerable gearbox lifetime enhancement progress has been made in recent years. According to multiple sources, today the reliability of geared and direct drive wind turbines is comparable.[7,25] Most wind turbine downtime is cited as being caused by relatively inexpensive 'dollar' components like broken fuses, faulty electrical connections, defects in print boards, etc.[26,27] Another fact frequently referred to in the wind industry is that power converters contain far more components compared to generators, increasing their failure rate.

6.8.1 Availability

Turbine availability is a function of reliability and service performance.

The German wind energy association Bundesverband Windenergie conducts annual service satisfaction surveys among German turbine owners, subdivided into different makes and models.

The only direct drive systems supplier in a range of surveys was Enercon, which for many years has occupied the top position. Enercon's EPK (in translation Enercon Partner concept) turbine upkeep concept is widely considered a major contributory factor to both the high customer service satisfaction rate and the excellent availability record of the turbines. EPK offers 97% technical availability for an operational period up to fifteen years, and the 'full service' example is increasingly followed by other major wind industry competitors.

6.8.2 Systems efficiency and rave earth elements (REEs) materials use

For a realistic view on power systems efficiency, it is important to consider individual main components' efficiencies as well as cumulative efficiencies.

Regarding electrical efficiency, a fast-speed PMG is known to offer superior partial-load efficiency compared to a DFIG of similar power rating. However, internal loss in a smaller-sized partial converter is less than in a full converter for similar generator power rating. Both factors should be taken into account when comparing a DFIG and PMG, whereas also systems integration serves as a key overall turbine systems efficiency enhancing feature.

Another principal factor impacting total system efficiency is turbine capacity. If, for instance, a given turbine operates with many full-load hours, the load point shifts to a range where the partial efficiency advantage of a PMG becomes far less prominent. There is a key and increasing wind industry trend for turbine models to be fitted with large rotors in combination with modest generator power rating.

In summary, while wind industry opinions continue to differ on total efficiencies and the benefits of different system solutions and their optimizing options, PMGs with full converter seem clearly to be on the increase. Also apparently winning at the expense of direct drive because of energy costs are two-stage medium-speed systems, but here again opinion is divided. It is also interesting to note that for a fast-speed PMG on average only about 10% of the magnet quantity is required compared to a direct drive PMG of similar power rating.[28] Finally, the future role for fast-speed geared including the use of DFIGs could prove more enduring than many in the industry would like others to believe.

6.9 References

1. de Vries, E. 'Offshore boost for Bonus; Danish wind firm is scaling-up its activities', *Renewable Energy World*, November–December 2004.
2. Kenetech 33M-VS product brochure, 1993.
3. de Vries, E. 'Windturbines zonder tandwielkast' (Wind turbines without gearbox) *Energietechniek*, February issue, No. 2, 1998.
4. Northern Power Systems, Verbal comment by Charles Vaughan, Vice President, Commercialization, at EWEA 2011,Brussels, Belgium.
5. 'Carbon fibre experts M Torres put up first prototype machine', *Windpower Monthly Magazine,* July 2001.
6. Sustainable Energy Technologies, Verbal comment by Gerald Hehenberger, CEO, at EWEA 2011, Brussels, Belgium.
7. Henk Polinder, 'Overview of and trends in wind turbine generator systems', at IEEE Power Engineering Society Annual Meeting, Detroit, 24–28 July 2011, pp. 1–8.
8. de Vries, E. 'Permanent solutions?' *Renewable Energy World*, March–April 2010.
9. Email correspondence with Peter Tavner. 8 May 2012.
10. de Vries, E. 'Turbine upgrade fuels offshore ambition', *Windpower Monthly*, February 2011.
11. de Vries, E. 'After 10 years GE goes back to DFIGs', *Windpower Monthly*, 9 October 2012.
12. de Vries, E. 'Multibrid; A new offshore wind turbine contender', *Renewable Energy World*, September–October 2004.
13. de Vries, E. 'Gamesa's new 4.5-MW wind market entry'. RenewableEnergyWorld.com, 23 November 2009.
14. Interview with Finn Strøm Madsen, President Vestas Technology R&D. 30 March 2011.
15. de Vries, E. 'Daewoo reveals 7MW offshore turbine', *Windpower Monthly* website, 13 December 2011.
16. de Vries, E. 'Permanent solution?' *Renewable Energy World*, March–April 2010.

17. de Vries, E. 'Innovation, the ingenious is always simple', *Renewable Energy World*, November–December 2007.
18. Verbal comments by Vensys Energy during a telephone enquiry, March 2011.
19. 'Close up – Northern Power plans 8MW offshore turbine', *Windpower Monthly*, 19 September 2011.
20. de Vries, E. 'Close up – sway turbines ST1o 10MW turbine', *Windpower Monthly*, 27 October 2012.
21. de Vries, E. 'Offshore boost for Bonus; Danish wind firm is scaling-up its activities', *Renewable Energy World*, November–December 2004.
22. Discussion with Henk Polinder, associate professor at Delft University of Technology in the field of electrical machines and drives.
23. de Vries, E. 'US firm launches novel CVT device', *Windpower Monthly*, 1 December 2011.
24. de Vries, E. 'Mitsubishi launches 7MW turbine', *Windpower Monthly*, 1 January 2012.
25. Discussion with Hansen Transmissions experts, September 2010.
26. NUON Presentation at Dutch scientific wind power conference, October 2010;
27. Verbal comments by a senior service technician of a major wind turbine supplier, November 2010.
28. Stephan Ebner and Teijo Kärnä, 'Challenges and solutions for the electrical drive train of large offshore wind turbines. Choosing the right technology and supply-chain partner', ABB presentation at Husum Wind Energy 2010, September 2010.

7

Case study of the permanent magnet direct drive generator in the Zephyros wind turbine

A. JASSAL, Delft University of Technology, The Netherlands,
K. VERSTEEGH, XEMC-Darwind, The Netherlands and
H. POLINDER, Delft University of Technology, The Netherlands

DOI: 10.1533/9780857097491.2.158

Abstract: The single-bearing permanent magnet direct drive generator of the Zephyros wind turbine has proven reliable in the field and is considered a successful design that has led to weight and cost reduction of the generator. Recent trends suggest the future of wind turbines lies in offshore installations, and Zephyros's generator was designed with these in mind. Much current research in the field of weight and cost optimization of wind turbines aims at putting more wind turbines into the overall energy mix.

Key words: Zephyros, generator design, permanent magnets, case study, operational experience.

7.1 Introduction

This chapter presents a case study of a permanent magnet direct drive (PMDD) generator used in a wind turbine named Zephyros. The generator is a multi-pole synchronous permanent magnet (PM) type which is fully integrated in the structural design as shown in Fig. 7.1. The turbine has three glass fibre reinforced plastic (GFRP) blades and a steel tubular tower. In order to make the design more attractive and to be able to extend the design life as much as possible, both the offshore and onshore markets were considered.

Zephyros is based on an earlier Lagerwey wind turbine of 750 kW power rating. The first turbine was installed in April 2002 and certification, design assessment and measurements have been completed.

The main focus of this chapter is the generator design; some important practical aspects are also discussed. The chapter starts with a discussion of site conditions and resulting design constraints, then the design process is described. Finally, some future trends and challenges are presented.

7.2 Design process and the resulting design

Since design considerations have already been covered in chapters 2 and 3, this chapter will focus on the design strategy chosen for the Zephyros.

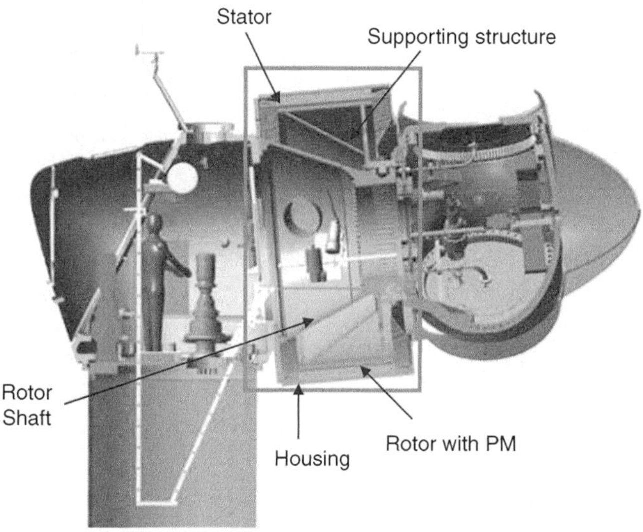

7.1 Z72/2000 wind turbine.

7.2.1 Site conditions and general layout

The foremost requirement for any wind turbine is the amount of wind available at a particular site: it is the first design constraint. The average availability of wind defines the amount of torque and speed possible at a location. For this particular generator the specified site conditions are:

1. Mean wind speed: 7.5–8.5 m/s at hub height.
2. Ambient temperature: 20–40°C.
3. General environment: clean, low salt content.

Location of generator in turbine: The generator can be located at the top of the tower, in front of or behind it (the usual position of a doubly fed induction generator with gearbox). The position depends on the type of generator system, mechanical design of the tower, size of the generator and cooling mechanism of the generator selected. Zephyros' generator has been placed in front of the tower as it leads to better cooling and better integration with the turbine hub, leading to weight reduction.

Bearing configuration: The bearing configuration depends on the mechanical torque, size and axial length of the generator, and load calculation of the turbine blades under various wind conditions. In general the configuration can be single-bearing or double-bearing. The Zephyros generator design is based on a single-bearing configuration coupled directly to the rotor

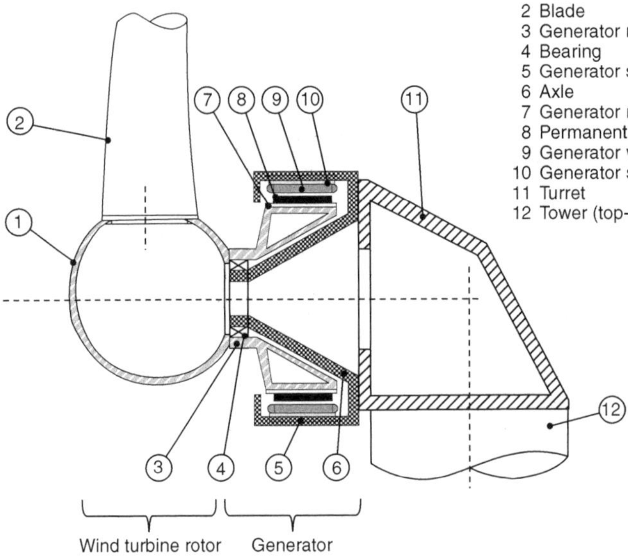

7.2 A typical single-bearing design as used in the Z72/2000 wind turbine.[1]

structure, eliminating the need for a long main shaft as used in traditional designs. This arrangement is shown in Fig. 7.2.

Although the bearing is of larger diameter, the overall weight is reduced because there is a very small shaft. A single multiple-row roller bearing was used, with the hub mounted on one side and generator carrier and rotor on the other. The stator is mounted on the opposite side of the carrier, which in turn is mounted on a compact casted nacelle frame.

Cooling: The generator is the totally enclosed type with IP54 enclosure. It is cooled passively from inside and by ambient air on the outside. This system falls within the IC40 cooling class.

7.2.2 Sizing and topology

The initial size and topology of machine is selected according to the following factors:

1. Heat dissipation required per unit surface area based on ambient temperature.
2. Transportation limits.
3. Force density required in a design (from experience).
4. Tip speed for an acceptable noise level.

The key design statistics for Zephyros are:

- Maximum outer generator diameter: 4 m.
- Nominal rotational speed: 18 rpm.
- Nominal power: 1500 kW.

Transportation constraints fixed the generator size at 4 m outside diameter. It was manufactured as a single assembly, as it is not safe to assemble a high-voltage machine in parts on top of a tower. The axial length was chosen to enable good heat dissipation from the whole outer surface. An inner rotor design with surface-mounted radially magnetized permanent magnets was selected. This choice reflects the thermal limits of the generator and possible saving in total volume, and hence cost, of the magnets. Based on the expected noise level, the nominal rotational speed of the generator was chosen as 18 rpm. However, a margin was allowed in the design for offshore operation, where the same generator can deliver 2000 kW at higher rotational speed and higher noise levels.

7.2.3 Selection of voltage level and converter

In a PMDD machine the field is always present, so that at rotational speeds greater than nominal, a higher voltage is always generated. The maximum voltage generated has to be deduced from the strength of field and rotational speed within the range of operation, allowing some margin for over-speed. This voltage level is calculated from the converter DC-link voltage which has to be used with the turbine. For the Zephyros generator, a 3 kV nominal voltage with 7.5 kV insulation level was chosen.[2] Choosing a high insulation level allows an increase in voltage level rather than current, should there be an increase in power output. Limiting the current ensures the copper losses and hence high heat generation are both limited, so that the constraint of maximum heat dissipation from the generator surface is not compromised.

A medium-voltage converter system with fewer components, rather than a low-voltage converter, was selected in order to achieve better efficiency. The AC–DC–AC converter (Fig. 7.3) in the tower base allows the generator to operate with a variable speed while the power is fed into the grid with a constant frequency of 50 Hz (or 60 Hz for countries where this applies). It also assures an average constant power output for wind speeds above rated. The power factor at the grid side is controllable at standstill as well as under operating conditions. Above rated wind speed, the blade pitch control maintains a more or less constant rotor speed till the speed overshoot margin. The rotor speed during power production is variable.

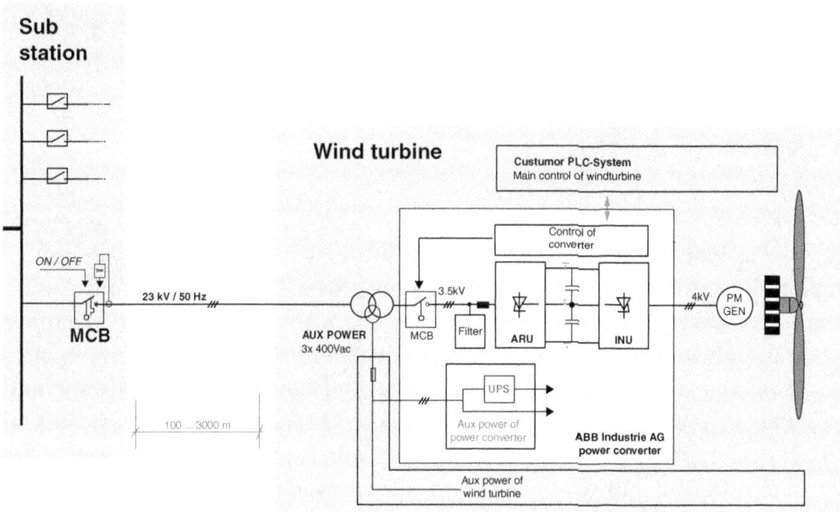

7.3 Single line diagram of the whole turbine.

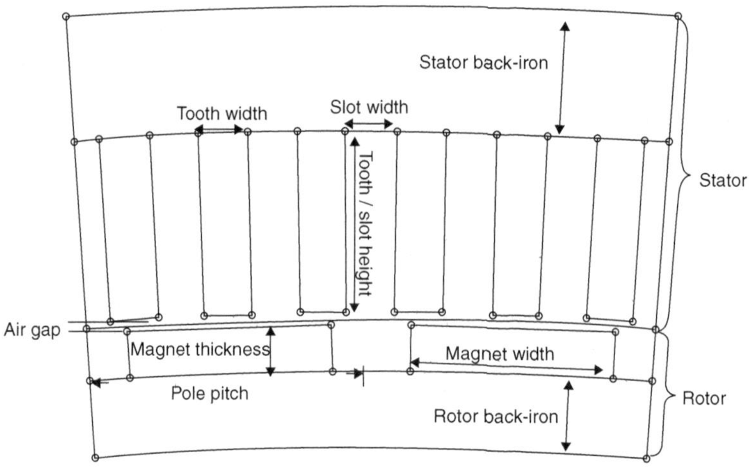

7.4 A general section of a radial-flux PM generator showing main dimensions.

7.2.4 Geometry and dimensions

After selection of the voltage level other dimensions of the generator stator are calculated. This is an iterative process and, using both experience and electromagnetic design equations, the dimensions of slots, teeth, stator and rotor back-iron and magnets are determined (Fig. 7.4).

Table 7.1 Design Features of Z72[1]

Rated shaft power	1670 kW	Temperature rise class	F
Rated electrical power	1562 kW	Insulation class	F(H)
Rated airgap torque	862 kNm	Standards	IEC 34
Rated voltage	3000 V	Protection by enclosure	IP54
Rated current	327 A	Cooling type	IC40
Power factor	0.92	Rotor inertia	35 000 kgm^2
Frequency	3–9.25 Hz (rated)	Total weight	47 200 kg
Rotational speed	9–18.5 rpm (rated)	Stator weight	25 000 kg
Number of poles	60	Rotor weight	12 500 kg
Pole angle	33.5°	Bearing support cone	5000 kg
Torque harmonics	100% fundamental (862 kNm)	Bearing weight	4000 kg
	<1% 6th harmonic 55.5 Hz	No. of PT100* temperature sensors installed in stator windings	6
	<1% 12th harmonic 111 Hz	PT100 sensors for generator air temperature monitoring	2
	<1% 24th harmonic 222 Hz	PT100 sensor for bearing temperature monitoring	2
Short-circuit current	569 A (sustained)	Air gap distance sensors	4
Ambient temperature	40°C	Bearing greasing unit	1
Radial pull	98 kN/mm between stator and rotor due to eccentricity	Maximum magnetic force	45 kN magnetic pulling force of one pole

*PT100: Platinum resistanceTemperature sensor with 100 Ω resistance at 0°C.

The resulting design summary is shown in Table 7.1.[2]

Based on electromagnetic design equations, the generator efficiency was estimated and is shown in Fig. 7.5.

7.2.5 Winding design

The winding scheme to be used in the generator is determined at an early stage, with options including single-layer or double-layer, distributed or concentrated type of winding. The winding choice is based on size and type of machine and the number of turns of the winding depends on the voltage to be generated by the machine (see Chapter 3 for more detail on windings).

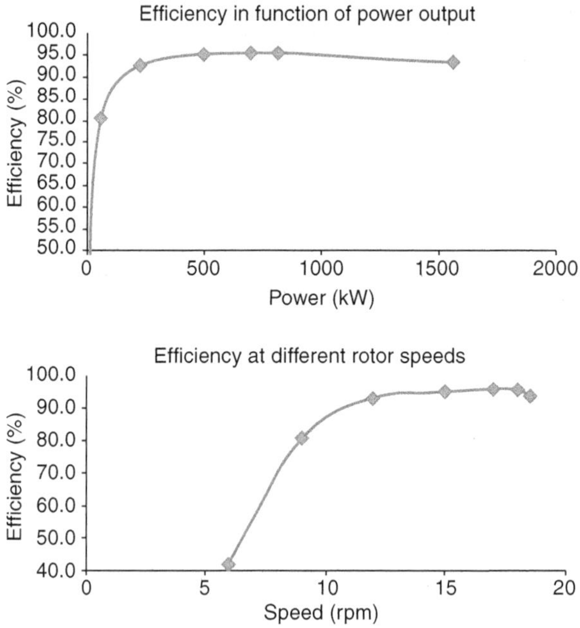

7.5 Generator efficiency curve.

For the Zephyros a full-pitch double-layer distributed winding was chosen, which had the following advantages:

1. Sinusoidal voltage output with low harmonic content.
2. The coils and end connections are symmetrical so just one type of coil is required.

Pre-formed or flat wire was used, which is appropriate for the chosen insulation level of 7.5 kV. Cheaper, mass-produced round wire could not be used because of its poor insulation quality. The slot-fill factor of 0.7 achieved with flat wire is better than that for round wire, typically 0.45, resulting in a weight saving in active material. The disadvantage of flat wire is the larger number of connections and the need for slot wedges. The stator for Zephyros is vacuum pressure impregnated which is also standard for medium-voltage machines.

7.3 Other design considerations

There are many other technical considerations for a successful design. In the case of large PMDD generators, the primary concerns are discussed briefly in the context of Zephyros.

7.3.1 Cogging torque

Cogging torque results from the interaction of magnets with the stator teeth. The PMs always try to align themselves in a position where total magnetic reluctance (resistance to magnetic flux set-up) is minimal. The main effects of cogging torque are audible noise, and torque ripple, which can cause fatigue in shaft especially if the shaft is long.[3] In an extreme case, cogging can be so high that the machine fails to start. In small machines the magnets are skewed to reduce cogging torque, but axial forces are produced, leading to stress in bearings and generator housing. For the Zephyros machine, cogging was dealt with by choosing the magnet span in such a way that it is minimal.[4,5]

7.3.2 Forces due to rotor eccentricity

The magnetic field in the airgap generates a force of attraction between the iron cored stator and rotor. If the rotor and stator are perfectly concentric, so that the physical airgap is uniform all the way round the machine, then all the attractive magnetic forces are balanced and there is no net force acting radially on rotor shaft. However, due to manufacturing tolerances there can be some eccentricity in the rotor position within the stator bore, which leads to a net force (also referred to as unbalanced magnetic pull) on the shaft. This was calculated for the Z72 and found to be about 98 kN/mm of eccentricity. This value is required for setting the shaft stiffness.

7.3.3 Short-circuit analysis

Short-circuit analysis is required in order to understand the impact of the resulting shaft torque and also the demagnetizing effects of the short-circuit current on permanent magnets. The magnets should be strong enough to counteract the demagnetizing short-circuit magneto-motive force (MMF) without getting permanently demagnetized. If the magnets are not strong enough, they can lose their magnetism (and hence utility) permanently. A forced short-circuit test was conducted and maximum short-circuit current of 569 A was recorded as against 327 A rated current. The demagnetizing MMF produced by this magnitude of short-circuit current was found to be easily negotiable by the magnets. More detailed measurements have been omitted because of commercial sensitivity.

7.4 Generator assembly

In the Zephyros, the generator structure forms part of the turbine's load-carrying capacity in combination with the single-bearing construction.

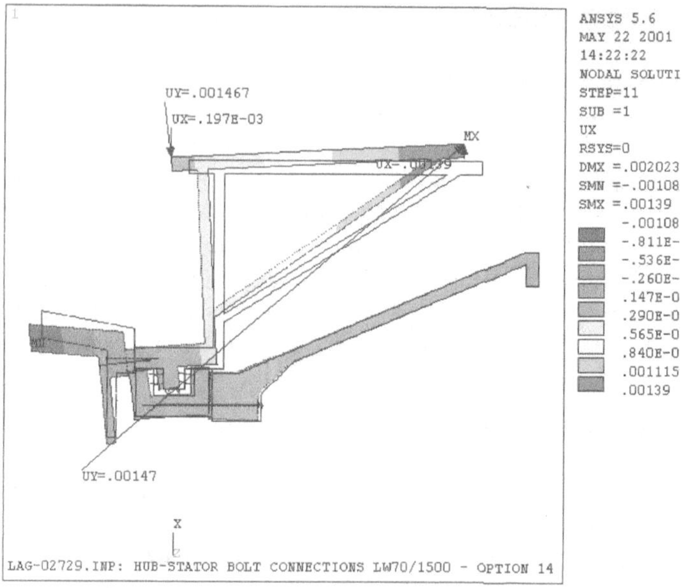

7.6 Finite element (FE) estimation of deflection of rotor structure.[2]

Finite element calculations have been performed to determine strength and stiffness of the structure and bolted connections. With a nominal airgap of 3 mm and an active material length of 1200 mm the requirements regarding deformation due to external wind and magnetic loads are challenging.

7.4.1 Generator housing and stator assembly

The generator housing is welded and the stator laminations are fixed into it using wedges.

7.4.2 Rotor and shaft assembly

The generator rotor and turbine hub form an integrated sub-assembly within the turbine. The rotor structure resembles a cup with support structure as shown in Fig. 7.1. A maximum deflection of about 2 mm has been calculated under extreme loads. The stator and rotor dimensions are determined more by the stiffness required to minimize deflection caused by the magnetic pulling forces rather than the material stresses. These calculations were performed by the ABB research centre in Mannheim, Germany. The result of a sample calculation is shown in Fig. 7.6.

7.4.3 Permanent magnets

For the magnet material neodymium iron boron (NdFeB) is used. The cost of high energy product (BH product) magnets fell to below 70–80 €/kg in 2005–2006, but increased by more than four times to about 300–350 €/kg during 2010–2011. The current price is about 100 €/kg. The magnets in the generator used for Zephyros are glued on steel modules and then magnetized. These modules are bolted on the rotor and a GRP (glass reinforced plastic) bandage is wrapped around the rotor before coating.

7.5 Generator testing

Many tests were performed on the generator in order to prove the design and establish performance characteristics. The back-to-back test set-up as shown in Fig. 7.7 was used. In back-to-back testing, two systems are mounted back-to-back, as drive equipment with low rpm and such high torque range are not readily available. These two machines are mechanically coupled, with one made to run as generator and the other acting as load. In the Zephyros back-to-back testing, the two machines were cooled with external fans. The speed and a position signal of the motor generator shaft were provided for over-speed tests and speed monitoring.

7.7 Back-to-back test set-up.

7.5.1 Insulation resistance

The stator's electrical insulation strength is tested by using an insulation tester or Megger. The values should be in the range of hundreds of mega ohms (>100 MΩ) and should increase with time when measured for longer periods (up to 10 min). The values for Z72 were excellent owing to vacuum impregnation.

7.5.2 Tests with converter and control

First the converter is checked for communication and then slowly each machine on the back-to-back test rig is rotated independently and if required the control parameters are adjusted. These checks were performed on Z72 and operation was found to be satisfactory. The main parameters which are checked by individual tests are:

1. Converter communication and control protocols.
2. Generated no-load voltage.
3. Back-to-back testing – generator on load and thermal test.

Power converter no. 1 on the left side of Fig. 7.8 is the device under test; the other one gives the load for the generator. Due to the losses in the converters, motor and generator, the output power of the converter on the generator side is reduced by about 15%. This resulted in a power of 1.5 MW from the generator side converter so that a test at maximum current could be executed. The following tests have been executed (limited to the generator):

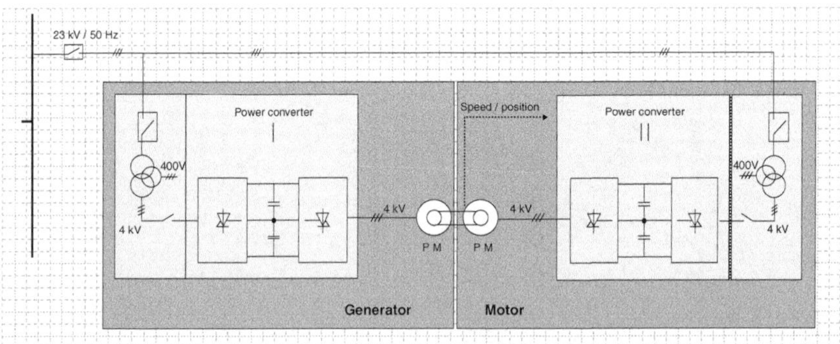

7.8 Back-to-back testing schematic.

1. *Light load test*: Each drive train is tested under its own protection on no load so as to check the normal function. Protection levels and functions are checked by forcing different faults on the converters.
2. *Over-speed test*: This test is to check the external over-speed protection.
3. *Forced generator short-circuit*: This test is to determine the short-circuit current and peak value of the torque.
4. *Firing through*: This test is to check the mechanical strength of the DC link and the braking capacity of the main circuit breaker.
5. *Continuous load test*: This test is to prove that the drive train meets the specifications in steady state conditions. During a single load test the power is ramped up to the maximum. The following parameters were monitored: the temperature in the generator, converter and main transformer, rotational speed, load angle, generator frequency, generator power, power of auxiliary systems, grid power output and total losses of test bench. The interaction between the generator and converter was also examined for over-voltages caused at generator terminals due to converter switching action (high rate of voltage change and reflection of travelling waves). The rated insulation voltage of generator should be such that it can withstand these over-voltages. For Zephyros, the insulation level was found to be sufficient.
6. *Earthing*: Optimization of the grounding concept and the flange filter to minimize the influence of voltage impulses (dv/dt) along with common-mode and differential-mode voltage to the generator.
7. *Control system response*: Dynamic variation of the torque to optimize the closed loop control.

The converter system and generator fulfilled their requirements under normal and extraordinary operating conditions. Measured values of the drive train were within its limits and the generator control system was able to track very closely the active power reference, demonstrating good dynamic performance. Generator reactive power was controllable in such a way that the terminal voltage did not exceed the rated value. However, the test had some limitations:

- The dynamic tests could not be carried out with nominal load. The reason for this is the very low inertia which is about 40 times smaller than that with the turbine rotor mounted.
- The cooling did not replicate real circumstances as no actual wind was present. The influence of switching on some external cooling fans could clearly be measured and provided reassurance. In practice, cooling can only be tested on site.

Other components tested included:

1. *Blades*: behaviour under fatigue and extreme loads.
2. *Pitch drive*: durability under endurance loads and thermal behaviour at extreme loads.
3. *Control cubicle*: vibration test, salt spray test and thermal test to meet specifications regarding corrosion, vibration and temperature range.
4. *Control software*: system and response test in the workshop to compare with the response of the computer model.

7.6 Operational experience and problems faced

The turbine was installed in April 2002 at Maasvlakte near Rotterdam, The Netherlands and became fully operational in November 2003. The generator and blades were mounted using an innovative hoist mounted on top of the tower as shown in Fig. 7.9. All the blades were assembled onto the hub at ground level and then the complete hub-rotor assembly was hoisted up. The generator was pre-assembled and mounted onto the tower in the same way before the hub-rotor assembly.

For the first year of operation the overall availability was 84%, which translates to a capacity factor of 27% for this site. If this is corrected to an expected availability of 97% it would result in 31% capacity, which is excellent for a site with an average wind speed at hub height of 7.5 m/s.

The measured power curve (as a function of the wind speed and measured according to IEC61400–12) given in Fig. 7.10 shows good comparison with the calculated curve and supports the excellent performance of the high-voltage generator-converter system.

However, some problems and limitations of the test set-up and operational experience are worth mentioning. These include:

7.9 Erection of Zephyros wind turbine.

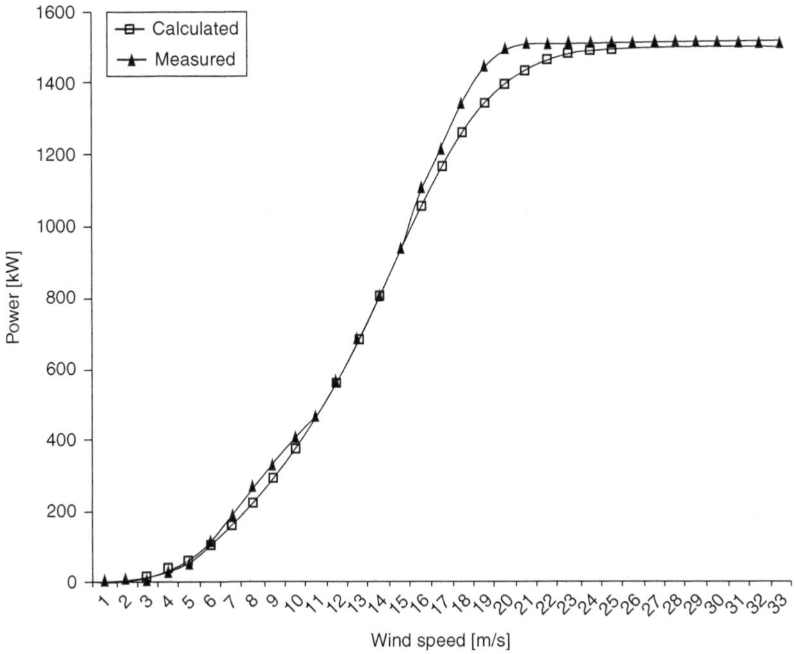

7.10 Calculated and measured power curve.

1. An initial problem was the noise generated due to resonance of the tower shell sections excited by the power converter switching frequency (480 Hz). This was solved by increasing the frequency to 800 Hz.
2. A heat run was performed for a period of 24 h continuously at full load. At an ambient temperature of 0°C the maximum stator temperature did not exceed 75°C. In practice, however, the maximum generator temperature measured during the first summer after installation was 86°C at an ambient temperature of 17°C. For cooling ventilation was applied for bearing and electronics in hub and nacelle.
3. Most of the other operational problems were in the components that had been adapted from the existing Lagerwey design. More attention should have been paid to these components during the scaling-up design process, but quality assurance (QA) aspects also contributed. The design of two components had to be adapted: wear and malfunctioning of the service brake, and control of the yaw brake passive torque to avoid overload on the yaw drives.

7.7 Reliability

The Z72 design process involved not only concept design, preliminary design and detailed engineering design, with shop drawings, specifications and design reports produced, but also specification of the so-called RAM (reliability, availability and maintainability) targets. In principle RAM targets should lead to predictable failure rates per main component or sub-system, providing some indication of system reliability. Details are omitted for reasons of commercial sensitivity.

In the early concept and design stages of a technical system it can be shown that the system will only really achieve the ultimate required availability goal some time towards the end of its specified lifetime. After the design stage is finished (and just before manufacturing starts), a considerable amount of the total lifetime is usually already over. At the same time most of the lifetime costs have already been locked in.

It is very clear that best improvement in availability at the lowest price can be achieved with higher returns before the design process is closed. This is also known as a 'reliability data sources management problem', which in the case of wind turbine design is hard to solve and requires extensive feedback from the organization concerned, especially the service department. To build up statistics the track record of a significant number of turbines is required. However, if this database is not available due to lack of operating experience, then the initial RAM targets are based on actual knowledge and experience, but they are subject to verification and adaptation when statistical data is available. From experience in the field, the generator system for Zephyros is highly reliable with no failures in generator system. However, minor power electronic problems have been reported, such as false alarms, communication errors, etc. Other problems were sensor-related, including malfunctioning of the temperature sensor due to a damaged cable, malfunctioning of the speed encoder and software control problems in pitching during the initial test run.

7.8 Future trends

The most significant recent trend in the field of wind energy generation is to a move offshore. There is also a trend to go for small numbers of large units rather than large numbers of small units due to maintenance issues. The same trend of a small number of large wind turbines can be seen in onshore installations.

In the light of increasing demand for renewable energy and the fact that wind is a major source of renewable energy,[6] an increasing amount of it is required in the overall energy mix. This is the main trigger for having more wind turbine (or wind park) installations in deeper sea areas. The main reasons for going offshore are:

- High availability of wind in offshore conditions and thus high energy yield.
- No noise restrictions.
- No visual impact.

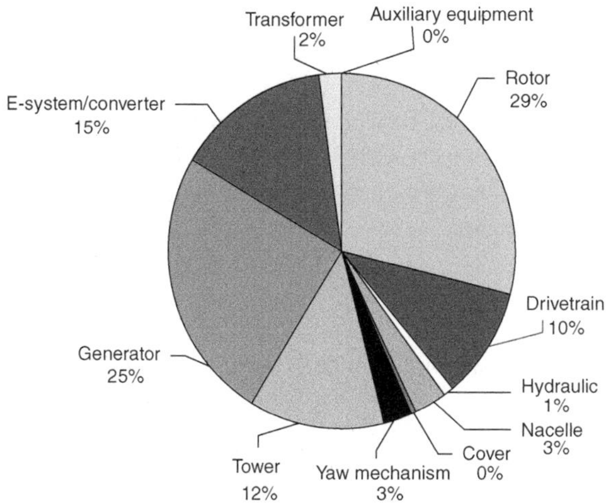

7.11 Cost breakdown of Zephyros turbine.

The Zephyros generator was also designed with offshore application in mind. In offshore conditions, at higher rotational speed, voltage and higher noise the same generator can deliver 2 MW of power to grid. The main challenges in this respect are:

- Limited access to turbines therefore maintenance should be minimal or reliability should be very high.
- High initial erection and commissioning costs.
- Environment is highly corrosive which demands a high level of protection from ambient conditions.

There has been a lot of interest in PMDD generators. Companies like GE and Siemens have entered into this topology with power ratings of 3–4 MW.[7,8] This type of generator has also been deemed suitable for offshore applications because of its reliability.

Wind turbines are huge structures and it is obvious that with the trend towards even bigger turbines in near future, erection and commissioning of such units will be challenging. Another important consideration is the input cost as it determines the price per unit of electricity. Consequently, much emphasis is also being placed on cost and weight reduction of wind turbines. The cost breakup for the Zephyros is shown in Fig. 7.11. The more expensive parts are main research areas for the weight/cost reduction.

7.9 Conclusion

The Zephyros wind turbine has been briefly introduced with a particular focus on the generator design and operational experience. It has proven reliable in the field and is considered a successful design. The innovative single-bearing design for such a size of wind turbine has led to weight and cost reductions. Recent trends suggest that the future of wind turbines lies in offshore installations, and Zephyros was designed with such application in mind. Research continues to be carried out in the field of weight and cost optimization of wind turbines which aims at putting more wind turbines in the overall energy source mix.

7.10 References

1. J.N. Stander, G. Venter and M.J. Kamper, 'Review of direct-drive radial flux wind turbine generator mechanical design', Research article published in *Wind Energy* Magazine, Issue 15, pp. 459–72, July 2011.
2. C. Versteegh, 'Design of the Zephyros Z72 wind turbine with emphasis on the direct drive PM generator', Presented at NORPIE, NTNU Trondheim, Norway, 14–16 June 2004.
3. J. Sopanen, V. Ruuskanen, J. Nerg and J. Pyrhonen, 'Dynamic torque analysis of a wind turbine drive train including a direct-driven permanent-magnet generator', *IEEE Transactions on Industrial Electronics*, vol. **58**, No. 9, pp. 3859–67, September 2011.
4. T. Li and G. Slemon, 'Reduction of cogging torque in permanent magnet motors', *IEEE Transactions on Magnetics*, vol. **24**, No. 6, pp. 2901–3, 1988.
5. Z.Q. Zhu and D. Howe, 'Influence of design parameters on cogging torquein permanent magnet machines', *IEEE Transactions on Energy Conversion*, vol. **15**, No. 4, pp. 407–12, 2000.
6. G. Shrestha, H. Polinder, D.J. Bang and J.A. Ferreira, 'Review of energy conversion system for large wind turbines', Proceedings of EWEC 2008, Brussels, 30 March–3 April 2008.
7. Siemens Press release no. ere201004062 http://www.siemens.com/press/en/pressrelease/?press=/en/pressrelease/2010/renewable_energy/ere201004062.htm
8. http://www.technologyreview.com/energy/23517/?a=f

8

Direct drive wave energy conversion systems: an introduction

M. PRADO and H. POLINDER,
Delft University of Technology, The Netherlands

DOI: 10.1533/9780857097491.2.175

Abstract: This chapter introduces the application of direct drive machines in wave energy conversion. Wave energy as a renewable resource is described, together with the principles and the challenges behind its capture and conversion. An overview of direct drive in wave energy is then presented, with a discussion of the current limitations and possible solutions.

Key words: linear generator systems, permanent magnet (PM) generators, ocean wave energy, direct drive.

8.1 Introduction

This chapter addresses the application of direct drive machines in wave energy conversion. Wave energy is a promising source of renewable energy, which man has been trying to harness for centuries (Clément *et al.*, 2002; Miller, 2003, 2004). However, it was not until the first oil shock in the early 1970s that interest in large-scale wave energy exploitation arose worldwide (Evans, 1981). In recent decades, remarkable progress has been made in the theory and experimental testing of wave energy and a myriad of different technologies has been developed around the world (Falcão, 2006). Although full-scale prototypes have already been tested at sea, proven fully functional commercial solutions are still not available. Most of the new technologies focus on electricity production, although other applications are also being developed, but at a much smaller scale, such as water desalinization (Folley and Cruz, 2008). Direct drive machines are not commonly found in wave energy applications, with only a few examples being built and tested in real marine conditions (e.g. Elwood *et al.*, 2007; Danielsson *et al.*, 2008; Prado, 2008).

This short introduction to wave energy describes world resources, and the principles and challenges behind the conversion of wave energy into electricity. A review of the history of direct drive machines in wave energy follows, and current state-of-the-art direct drive machines are discussed, identifying problems and indicating directions for further research.

8.2 Wave energy

Wave energy is a promising but also quite a challenging form of renewable energy to tap. This section will discuss the world resources, the principles of its conversion into electricity and the challenges behind the design of wave energy converters.

8.2.1 World resources

The sun supplies our planet with a very generous 122 000 TW continuous flow of electromagnetic energy in the visible range (Ahrens, 1995). This solar energy drives many important processes on the surface of the planet before being reradiated back into space as infra-red radiation. The water cycle, the heated surface of the oceans and lithosphere, all photosynthetic biomass and the wind are examples of direct manifestations of the stream of solar energy that crosses our planet. Wave energy is, however, an indirect expression of the solar energy stream since it results from the action of winds blowing over the oceans. Waves propagate on the ocean surface until they reach the shore, where their remaining mechanical energy is dissipated as heat. The total wave resource reaching the world's shoreline is estimated to be approximately 2 TW, similar to total world electricity production (Barstow *et al.*, 2008). However, the wave resource is not evenly distributed around the planet. It is more concentrated at mid-latitudes where in general the wind resource is also stronger (see Fig. 8.1).

Wave energy as a resource has both positive and negative aspects. Some positive aspects are:

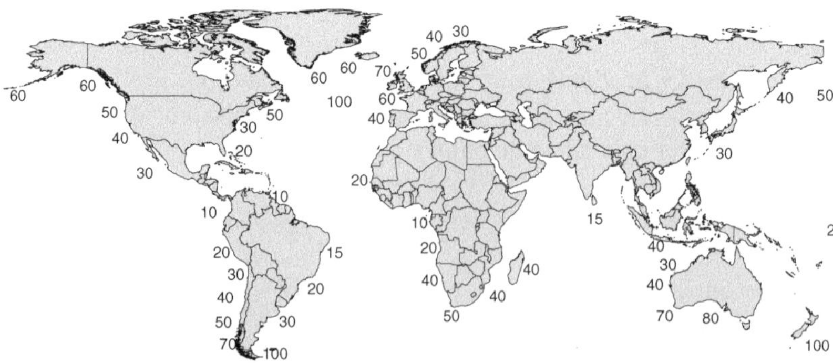

8.1 World distribution of offshore wave power – annual average (kW/m).

1. Higher density of energy compared to wind or solar energy, which allows more compact devices for the same power levels.
2. Wave intensities can be predicted some days in advance, which assists with grid management.
3. Wave energy production occurs along shorelines, where most urban centres are located.
4. The wave resource is usually stronger during winter time (Barstow *et al.*, 2008), when energy demand tends to be also higher.

Negative aspects of wave energy include:

1. The sea is an aggressive environment (mechanically and chemically).
2. Waves are highly random both in height and period, which implies that wave energy converters (WECs) will have to work continuously at non-steady operating conditions.
3. Due to its oscillating nature, the mechanical energy of ocean waves is not transmitted continuously but by pulses with twice the frequency of the waves.
4. Offshore accessibility is more restricted, which will condition installation and maintenance operations of WECs.

8.2.2 Fundamentals of wave energy conversion

From the electrical power engineering point of view, a WEC can be seen as an AC/AC converter, which converts a mechanical power signal with variable force and frequency into an electrical power signal with constant voltage and frequency (see Fig. 8.2).

In general, the process of wave energy conversion into electricity requires several steps. The first step is energy capture from the wave into the WEC

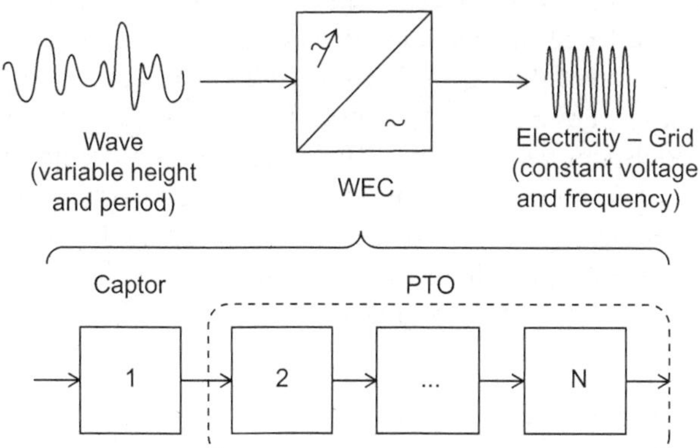

8.2 Wave energy conversion as an AC/AC converter and its basic functional units (captor and PTO).

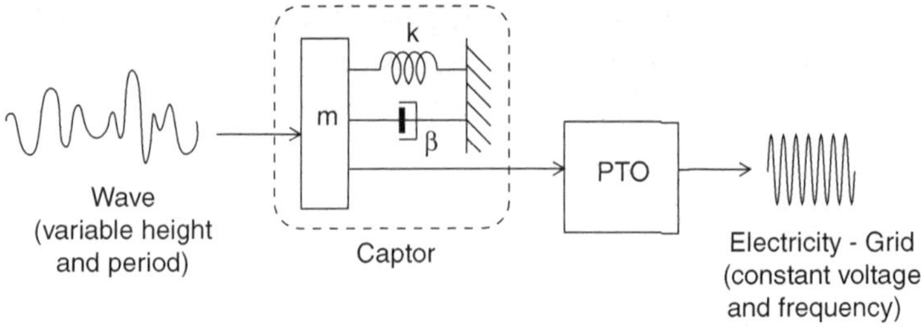

8.3 One degree of freedom oscillating captor (mass-spring-damping model).

and is done by the captor, while all subsequent energy conversions are done by the power take-off (PTO) unit. The captor therefore provides the interface of the WEC with the sea and it generally consists of a mechanical system with one or more degrees of freedom of motion (linear or rotational) excited by the waves. Different examples of captors are rigid bodies, overtopping devices, oscillating water columns and flexible bodies (Graw, 1995; EMEC, 2011; WAVEC, 2011). The function of the PTO is to extract the mechanical energy stored temporarily at the captor and convert it into electricity. The PTO may have several stages of energy conversion, moving, for example, through hydraulic, pneumatic and mechanical to electrical (Salter *et al.*, 2002; Curran *et al.*, 2008); or may directly convert the mechanical energy from the captor into electricity, using electromagnetic machines (Polinder *et al.*, 2005), electroactive materials (Prahlad *et al.*, 2005) or magnetohydrodynamics (SARA, 2011).

In general there will be a rectifying step at a certain point along the energy conversion chain of the WEC, as occurs with AC/AC converters with DC links. Depending on the WEC technology, the rectification process can occur directly at the captor or, more commonly, at the PTO. Examples of rectifying captors are overtopping devices (Tedd *et al.*, 2008), submerged horizontal flat plate (Graw, 1995) or Darrieus turbine (Scheijgrond *et al.*, 2006). In these cases, the motion of the captor is unidirectional, not oscillating like the waves. The majority of WEC concepts, however, use oscillating captors, for which linear models can be developed (Falnes, 2002). An oscillating captor with one degree of freedom can be modeled as a mass spring system with frequency ω, dependent inertia m, stiffness k and damping β (see Fig. 8.3).

By using the concept of mechanical impedance Z (ratio of force complex amplitude F over velocity complex amplitude v),

$$\frac{F(\omega)}{v(\omega)} = Z(\omega) = \underbrace{R(\omega)}_{\text{Resistance}} + i\underbrace{X(\omega)}_{\text{Reactance}}$$

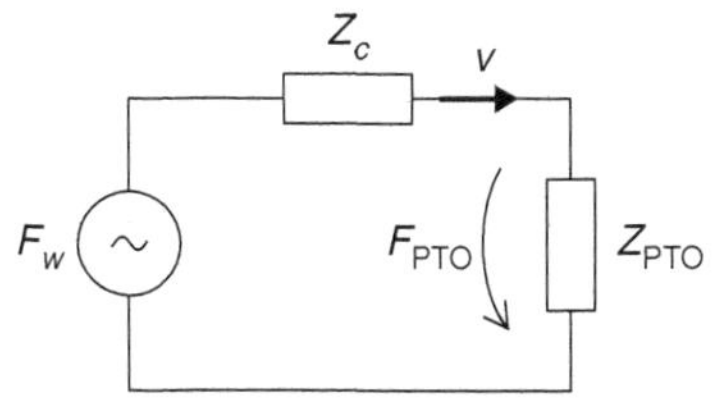

8.4 The equivalent electrical circuit of the captor's dynamics.

it is possible to describe the dynamics of the captor using an analogue electrical circuit as illustrated in Fig. 8.4, where:

- F_W is the complex amplitude of the wave force acting on the captor;
- F_{PTO} is complex amplitude of the PTO force acting on the captor;
- V is the complex amplitude of the captor's and PTO velocity;
- Z_C is the captor's mechanical impedance given by

$$Z_c(\omega) = \underbrace{\beta(\omega)}_{R_c(\omega)} + \underbrace{im(\omega)\omega - \frac{k(\omega)}{\omega}}_{X_c(\omega)};$$

- Z_{PTO} is the PTO mechanical impedance given by

$$Z_{\text{PTO}}(\omega) = R_{\text{PTO}}(\omega) + iX_{\text{PTO}}(\omega).$$

At frequency ω, the complex amplitude of the PTO's velocity is given by

$$V = \frac{F_W}{Z_C + Z_{\text{PTO}}},$$

and the PTO input apparent mechanical power is

$$P_S = \frac{1}{2} F_{\text{PTO}} V^* = \frac{1}{2} Z_{\text{PTO}} |V|^2,$$

with its active and reactive components given by

$$P_A = \text{Re}\{P_S\} = \frac{1}{2} R_{\text{PTO}} |V|^2,$$

$$P_R = \text{Im}\{P_S\} = \frac{1}{2} X_{\text{PTO}} |V|^2.$$

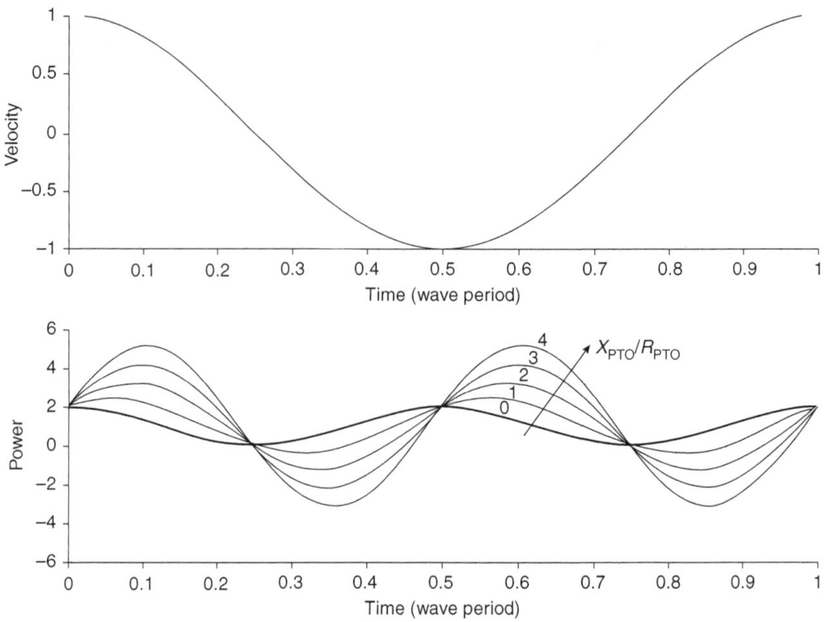

8.5 PTO velocity and input mechanical power signals for different mechanical impedances (constant resistance and increasing reactance). Pure input active power signal is depicted with a thicker line.

The active power P_A corresponds to the average mechanical power absorbed by the PTO from the captor, while the reactive power P_R corresponds to the zero average energy flow back and forth between the captor and the PTO during each wave cycle. Thus when Z_{PTO} is purely resistive, the PTO works only as a generator extracting energy from the captor, while when Z_{PTO} also has a reactive component, the PTO will work both as a generator and a motor. The PTO input mechanical power signal has an average value equal to the active power and an AC component at 2ω equal to the apparent power, as depicted in Fig. 8.5.

In general, the PTO energy losses increase with the power levels that it has to handle. Therefore as the input mechanical reactive power of the PTO increases, its average power conversion losses increase, as well as its rated power, resulting in a more costly and less efficient system.

Optimal power conversion

From Fig. 8.4, it is clear that maximum power transfer from wave to the PTO happens when the PTO's mechanical impedance is matched to the captor's mechanical impedance, that is

$$Z_{\mathrm{PTO}} = Z_C^* \Leftrightarrow \begin{cases} R_{\mathrm{PTO}} = R_C \\ X_{\mathrm{PTO}} = -X_C \end{cases},$$

which corresponds to the maximum PTO input active power

$$P_A^{\mathrm{OPT}} = \frac{1}{8}\frac{|F_W|^2}{R_C}.$$

This optimum power absorption comes however at a cost of having an associated PTO input reactive power equal to

$$P_R^{OPT} = -\frac{X_C}{R_C} P_A^{OPT}.$$

The higher the reactance of the captor, the smaller the PTO conversion efficiency and therefore the smaller the power output of the PTO. On the other hand, if the PTO mechanical impedance is constrained to be purely resistive, then the maximum power transfer from wave to PTO occurs when

$$Z_{\mathrm{PTO}} = |Z_C| \Leftrightarrow \begin{cases} R_{\mathrm{PTO}} = \sqrt{R_C^2 + X_C^2} \\ X_{\mathrm{PTO}} = 0 \end{cases}.$$

In this situation, there will be no PTO input reactive power, so PTO power conversion efficiency will be highest but the input active power will be smaller than the optimal P_A^{OPT}. Figure 8.6 depicts how the input active power of PTO degrades, when the PTO mechanical impedance deviates from the optimal case.

In general, the maximum power conversion does not coincide with maximum power absorption and it will depend on the PTO's efficiency and the amount of reactive power it has to sustain in order to increase power absorption.

To achieve high levels of wave power conversion it is therefore important to have (a) an efficient PTO, and (b) a small captor's reactance in the range of the ocean wave periods. The first requirement can be achieved by the use of direct drive machines. The second requirement is a power factor correction of the captor by mechanical means and can be achieved by (a) reducing the captor's inertia m and stiffness k, and (b) adjusting them relative to each other, such that

$$X_{\mathrm{Captor}}(\omega) = 0 \Leftrightarrow \omega = \sqrt{\frac{k}{m}}.$$

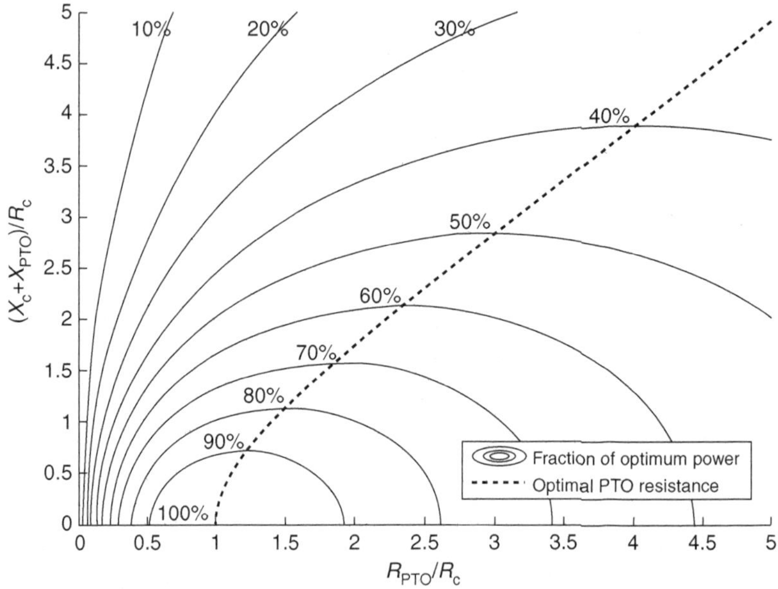

8.6 Fraction of optimal input active power absorbed by the PTO, function of its mechanical impedance matching.

Usually it is only possible to satisfy the above condition for a particular frequency, the so-called natural frequency. Matching the natural frequency of the captor to the mean frequency of the incoming waves is the fundamental operating principle behind most WECs, as it is in radio transmission when we tune a radio receiver to the base frequency of a radio station.

Power output smoothing

To minimize the power fluctuations transmitted to the grid, different solutions are possible:

- incorporate in the PTO an energy storage system to filter out the power oscillations;
- place several captors along the wave propagation direction, so the power oscillations from the different captors can be cancelled out due to their relative phase shift.

Examples of different WEC technologies using one or both of the above-mentioned solutions for power output smoothing are shown in Figs 8.7–8.9.

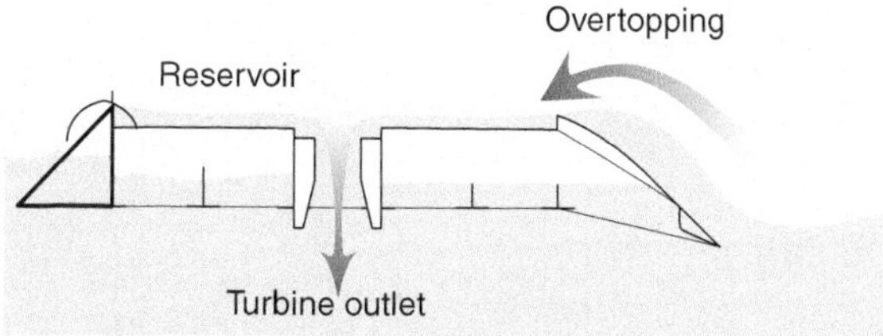

8.7 Example of WEC with internal energy storage (water reservoir) – The Wave Dragon (Wavedragon, 2011).

8.8 Example of WEC with multiple captors distributed along wave propagation direction – The Wave Star (Wavestar, 2011).

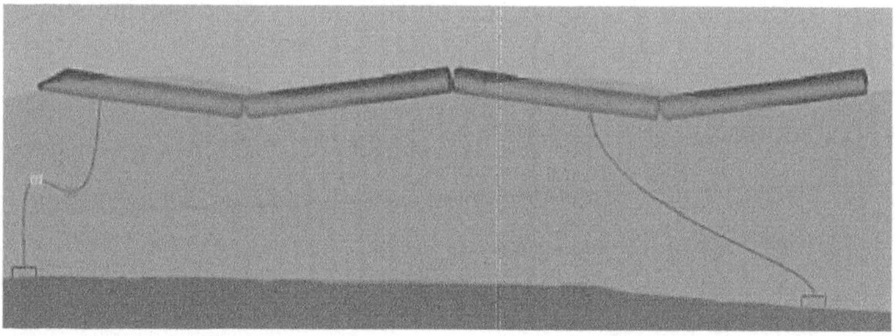

8.9 Example of WEC with multiple captors distributed along wave propagation direction and internal energy storage (each joint is equipped with an hydraulic PTO plus an accumulator) – The Pelamis (Pelamis, 2011).

8.2.3 Design challenges in wave energy conversion

A WEC presents various challenges in different aspects of its design.

1. Power conversion chain

   - High efficiency for a wide range of different operating conditions. This requires mechanical systems to be designed with large bandwidth and adjustable natural periods.

- Highly fluctuating input power signals (peak power is typically ten times the average power during a steady sea state).
- Keeping linear motions within strokes limits, without strong impact loads.

2. Structure and mooring

- Ability to survive extreme events (storms).
- High random mechanical loading, with approximately 5 million load cycles/year.

3. Installation and maintenance

- Easy mob/demob operations. Complicated installation or removal of the devices may imply very limited annual time windows due to weather constraints.
- Easy access to device or its more critical components.

4. Reliability

- High reliability of device components and assembly. Low reliability has a dramatic impact on the device's downtime since offshore access is difficult and costly.

5. Economics

- Cost-effectiveness of the device. It has to produce enough energy to pay back during its lifetime all the costs of construction, installation, operation, maintenance and decommissioning, as well as providing a profit margin.

8.3 Direct drive in wave energy

Direct drive PTOs are an appealing option for wave energy conversion. A historical review will be given, together with the description of the current state-of-the-art. Present limitations to overcome and possible solutions will be discussed.

8.3.1 Introduction

The use of direct drive machines for the PTO in WECs has positive effect in two very important respects:

- Efficiency, due to a reduced number of energy conversion steps.
- Reliability, due to a reduced number of components.

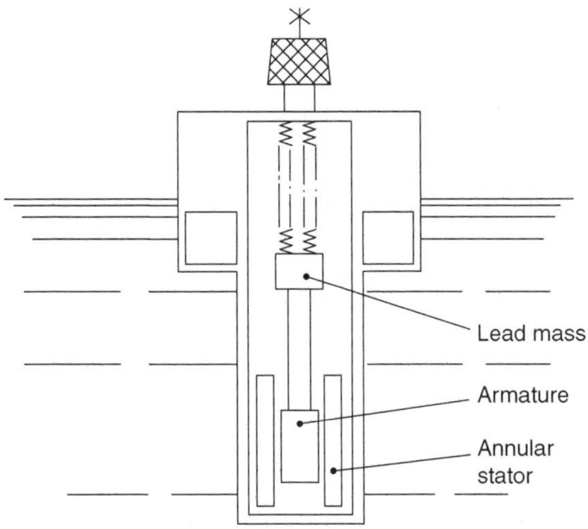

8.10 Diagram of a patent of a linear generator application in wave energy, 1970 (Leishman and Scobie, 1976).

For this reason, direct drive PTOs have been an appealing solution for wave energy developers. However wave energy conversion creates very demanding conditions for direct drive machines:

- irregular motion, with variable amplitude and frequency (0.1–0.2 Hz);
- high forces and small velocities (in the order of 1 m/s);
- need for cost-effectiveness.

The first specification requires the use of an AC/AC converter to allow grid connection of the electrical machine. The second and third specifications are directly related to the electrical machine (its topology and materials). Complying simultaneously with these two specifications presents a design challenge, as electrical machines favor high velocities. This has been the main reason for the limited number of direct drive machines applied to wave energy conversion and why WECs with higher captor velocities are usually preferred for direct drive PTOs.

8.3.2 Historical review

Patents for direct drive machines in wave energy can be traced back to 1970, with British Patent No.1316950 describing a device invented at the University College of North Wales (Leishman and Scobie, 1976). It consisted of a closed buoy equipped with an annular PM linear generator driven by the internal inertial force acting on the translator (see Fig. 8.10).

This concept was taken up and developed further by a UK company which built and tested a small prototype. The project was later abandoned

8.11 The 'Duck' WEC concept of the Edinburgh Wave Group – view of an array and transverse section (EWG, 2011).

after sea trials and numerical simulations showed that even for an average power output of 1 W, the design required a very large mass and presented serious mechanical difficulties.

Also in the early 1970s, one of the leading wave energy developers, the Edinburgh University Wave Power Group, analyzed the option of a direct drive PTO for their famous device 'The Duck' (see Fig. 8.11).

The Edinburgh concept consisted of a series of nodding bodies ('ducks'), moving around a fixed common central spine. The direct drive option was based on a PM rotary generator between each duck and the spine. At that time, the best PMs available were barium ferrite (with a remanent flux density of 0.35 T), which resulted in a design with an excessive mass of iron and copper. In addition to the prohibitive cost of the machine, its weight was a problem (some smaller ducks would have sunk simply under the weight of the generator). Since solid state inverters were still in their infancy and an expensive technology, the grid connection of the PM generator was an additional problem to solve (Salter *et al.*, 2002). The direct drive solution was therefore dropped by the Edinburgh team, which ended up developing a novel and sophisticated hydraulic PTO solution for 'The Duck' (Taylor, 2008).

Since then important advances have been made in the development of both PMs and power electronics. The first NdFeB magnets became commercially available in 1983 (Gieras, 2008) and by the early 1990s, thanks to advances in power electronics, AC drive systems replaced the DC drive systems in almost all application fields (Yano *et al.*, 2004). During the 1990s, the cost of both NdFeB magnets and power electronics saw a tenfold decrease over approximately a decade.

These events have led to renewed interest in direct drive machines for wave energy conversion during the last decade. In 1999, Direct Thrust Designs Ltd designed and built a sub-kW prototype linear PM machine applied to a WEC (Kelly, 2000) and tested it on the UK's North Devon coast (see Fig. 8.12).

The machine had tubular geometry, with the magnets attached to a buoy that would slide along the stator, which was rigidly connected to the sea bed. The unit was designed to generate near-mains voltage at mains frequency,

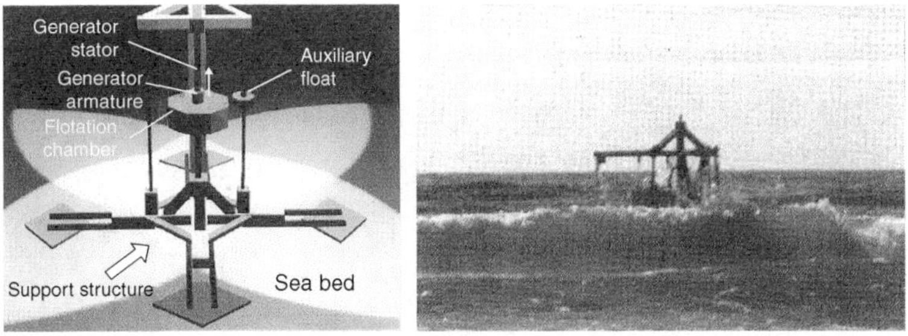

8.12 Experimental linear generator designed and built by DirectThrust Designs Ltd at the UK in 1999 – diagram (left) and photo during sea trials (right) (Kelly, 2000).

given the predominant wave conditions. An array of domestic light bulbs was lit onshore, through a power cable connected to the generator.

A number of offshore tests of larger-scale WECs equipped with direct drive PTOs were carried out just a few years later:

- 2004 – AWS 2 MW full-scale prototype in Portugal (see Chapter 9).
- 2006 – Uppsala University 10 kW full-scale prototype in Sweden (Danielsson *et al.*, 2008).
- 2007 – Oregon State University 10 kW prototype in the USA (Elwood *et al.*, 2007).

Figure 8.13 illustrates, for each device, the WEC principle, its linear generator's geometry and the tested prototype.

Note the large size of the machines compared, for example, to a standard electrical machine with an equivalent power level. All three direct drive machines can be considered state-of-the-art in wave energy and are based on the same type of electrical machine (topology and materials):

- Motion: linear;
- Type: longitudinal flux machine (LFM);
- Translator: NdFeB PMs + iron yoke;
- Stator: copper windings + iron core.

There are some differences between specific design features such as cooling, generator geometry, bearings, lubrication and construction details.

A rotary PM LFM has also recently been considered in the WEC SEAREV (Ruellan, 2007) (see Fig. 8.14).

The SEAREV is a rotational version of the wave energy concept shown in Fig. 8.10 and consists of an enclosed buoy with an internal pendulum that swings with the motion of the waves. From the relative motion between the

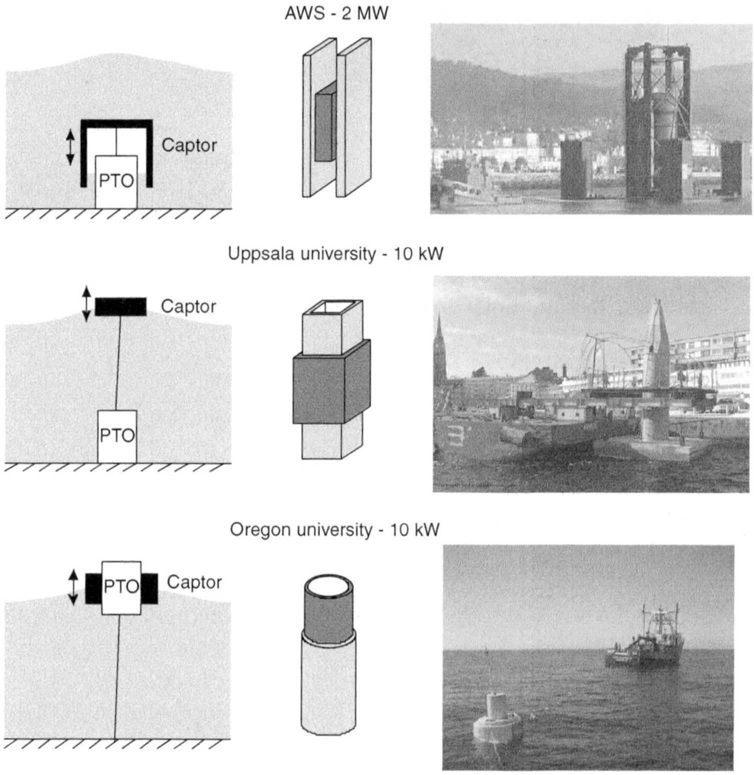

8.13 Diagrams and photos of the WECs equipped with direct drive PTO that have been tested offshore in recent years (Prado, 2008; Oregon, 2011; Uppsala, 2011).

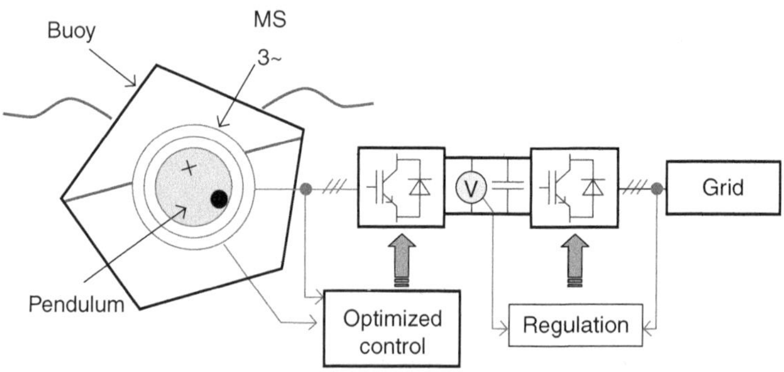

8.14 Direct drive option for the SEAREV WEC (Ruellan, 2007).

pendulum and the buoy, mechanical energy can be converted into electricity through a direct drive rotary machine. The design methodology of the generator was based on the minimization of costs (total losses or total cost of the active parts) with genetic algorithms (Ruellan, 2007). A sensitivity analysis of the design was performed and the results showed that the total cost of active parts was minimized when:

- NdFeB magnets are used instead of ferrites;
- laminated silicon steel is used instead of iron powder;
- airgap is minimized;
- cooling is maximized.

However, to date neither SEAREV nor other WECs with rotary direct drive machines have been built and deployed at sea.

8.3.3 Current limitations and possible solutions

Despite the above-mentioned potential benefits of direct drive machines in wave energy, the current state-of-the-art linear generators have still some negative characteristics (Polinder *et al.*, 2007):

- **Large size**, compared to standard electrical machines of similar power levels. This is a consequence of the low speeds typical in wave energy (in the order of 1 m/s), compared to the airgap speeds of standard higher-power electrical machines (above 25 m/s). For the same power level, smaller velocities therefore require higher forces. Since the force density of the electrical machines is limited (due to cooling and iron saturation), their size will necessarily be larger.
- **Large translator–stator attractive forces**, due to the magnetic interaction between the translator's PMs and stator's iron. For a generator with a perfectly symmetric geometry (cylindrical or double-sided) the translator–stator attractive forces would cancel out exactly. However, any small asymmetries in the machine (due, for example, to construction tolerances, bearing wear, thermal expansion or mechanical deformation) will cause an almost exponentially growing net attractive translator–stator force on the side where the airgap is smallest. This large unstable net attractive force therefore imposes demanding requirements on the mechanical design of the machine, in particular on the generator's support structure and bearing system.
- **Limited efficiency**, due to low velocities. At low velocities, the generator efficiency is limited mainly by the resistive losses. This is because the induced electrical field in the generator's conductors drops with speed but the electrical field necessary to overcome the resistivity of the conductors

does not. Efficiency can be improved by increasing the section of the conductors but the cost and size of the machine will also increase.

- **Large airgap** resulting directly from the mechanical design of the machine, which has to accommodate (a) the manufacturing tolerances (which increase with the size of the machine); (b) the deformations of the machine due to mechanical loading and thermal expansion; and (c) the wear on the bearings. The larger the airgap, the smaller the magnetic flux density and therefore the larger the machine size.
- **Partial translator–stator overlap**, which results in active material not being fully used during the operation of the machine. To generate maximum force for a larger fraction of the stroke, translators are made longer than the stators, since it is cheaper than making the stators longer. This problem is, however, specific to linear generators and is not present in rotary machines, where rotor and stator have a constant full overlap.

All these technical characteristics, in particular the large size, have a negative impact on the machine's cost. There are a number of possible solutions to these problems but further research is required to determine how practical and cost-effective they are (Polinder *et al.*, 2007):

1. Increase the speed of the machines
   This could be done by:

   - Introducing a gearbox between the captor and the PTO. However this poses challenges in the mechanical design due to the high and oscillating mechanical loads that the gearbox has to cope with.
   - Increasing the captor's stroke and minimizing its mechanical reactance for the range of expected wave frequencies. Although higher captor speeds are appealing from the electrical point of view, they also bring problems from the mechanical point of view, such as stronger non-linear hydrodynamic effects (e.g. drag losses, which grow with the cube of velocity).

2. Different machine topologies
   Various topologies have been proposed in recent years (Mueller *et al.*, 2000; Mueller, 2002; Baker *et al.*, 2004; Mueller and Baker, 2005; Danielsson *et al.*, 2008) to tackle the following specific aspects:

   - Reduce machine size by using variable reluctance machines, in particular transverse flux machines (TFMs). These machines are known for having higher force densities and efficiencies than LFMs but requiresmall airgaps and a more complex construction, and exhibit small power factors and complicated three-dimensional (3D) flux paths.

- Reduce translator–stator forces by using air-cored machines. By eliminating the iron from the stator, not only are the large attraction forces between translator magnets and iron stator eliminated, but also higher-power factors are obtained. However, the force density of the machine is also reduced due to increased flux leakage.

3. New materials
High temperature superconductors (HTSs) are perhaps the most promising new materials that could improve significantly the design of the magnetic and electrical circuits, in terms of both size reduction and increased efficiency. HTSs could be applied in the generator design to:

   - Generate stronger magnetic fields than is possible with current PMs, with modest power requirements. This would allow the generator's force density to be increased and also provide opportunites for air-cored topologies. Studies and experimental work done on synchronous machines with HTS rotor field excitation show several advantages compared to conventional PM synchronous machines with power levels above 1 MW, such as a weight reduction of approximately 50%, increased efficiency over the full operating range and a several times increase of overload capacity (Maki *et al.*, 2005; Gieras, 2008).
   - Eliminate resistive losses in the stator windings, increasing efficiency and enabling a higher electrical loading.

A comparison of different machines topologies for wave energy application was performed by Polinder *et al.* (2005), using the AWS WEC as a reference case (maximum 1 MN force, 2.2 m/s velocity and 7 m stroke). The same base materials were considered for each topology (NdFeB magnets, laminated silicon steel and copper windings) and the comparison criteria were the total cost of the active materials of the generator plus the cost of its energy losses during 5 years of operation. The results are shown in Fig. 8.15.

As Fig. 8.15 shows, only PM TFM performs slightly better than the conventional PM LFM, with an almost identical active materials cost but with an energy loss reduction of approximately 25%. Taking into account that power electronics and construction costs are expected to be higher in TFMs, no significant improvements can therefore be anticipated with this topology compared to PM LFMs. It should be also emphasized that the costs of the supporting structure of the generator and its bearing system are not included in the results shown in Fig. 8.15, and this may be a factor in the larger cost difference between the air-cored and iron-cored machines. These results may help to explain to some extent why all current state-of-the-art direct drive machines in wave energy conversion are based on conventional PM LFMs.

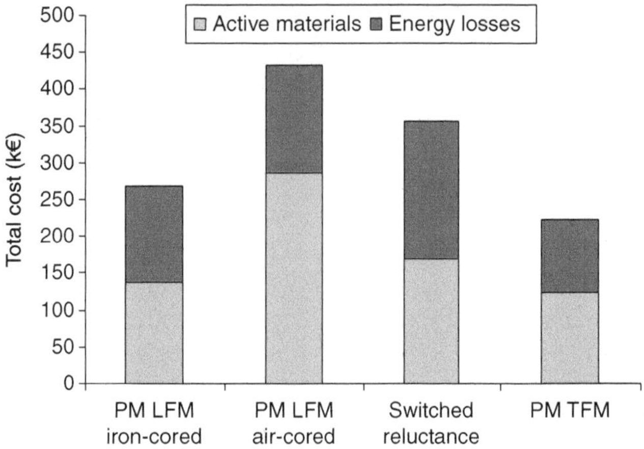

8.15 Total cost of a linear generator for the AWS WEC, considering different machine topologies (Polinder *et al.*, 2005).

8.4 Conclusion

This chapter has addressed the application of direct drive machines in wave energy conversion.

The highly irregular motions with small velocities and high forces that characterize wave energy conversion make the design of a cost-effective direct drive machine a real challenge. Although developers seem to show a preference for the conventional PM LFM topology, the large dimensions and bearing loads characteristic of all machines have a negative impact on their cost. At the present time, no alternative topology is able to bring significant improvements to the overall design of the generators. However, step changes in the cost and performance of machines may be possible if new materials such as HTSs are considered in the design.

8.5 Acknowledgement

This work was supported in part by a Marie Curie Initial Training Network of the European Community's Seventh Framework Program under contract number PITN-GA-2008-215414, the Wavetrain2 program.

8.6 References

Ahrens TJ (1995), '*Global Earth Physics – A Handbook of Physical Constants*', AGU.

Baker NJ, Mueller MA and Spooner E (2004), 'Permanent magnet air-cored tubular linear generator for marine energy converters', *2nd International Conference on Power Electronics, Machines and Drives*, **2**, 862–7.

Barstow S, Mollison D and Cruz J (2008), 'The Wave Energy Resource', in Cruz J, *Ocean Wave Energy – Current Status and Future Perspectives*, Springer-Verlag, 93–132.

Clément A, McCullen P, Falcão A, Fiorentino A, Gardner F, Hammarlund K, Lemonis G, Lewis T, Nielsen K, Petroncini S, Teresa Pontes M, Schild P, Sjöström BO, Sørensen HC and Thorpe T (2002), 'Wave energy in Europe: current status and perspectives', *Renewable and Sustainable Energy Reviews*, **6**, 405–31.

Curran R, Folley M, Danielsson O, Thorburn K, Leijon M, Taylor J and Cruz J (2008), 'Power take off-systems', in Cruz J, *Ocean Wave Energy – Current Status and Future Perspectives*, Springer-Verlag, 189–285.

Danielsson O, Thorburn K and Leijon M (2008), 'Power take off-systems (direct drive – linear generators)', in Cruz J, *Ocean Wave Energy – Current Status and Future Perspectives*, Springer-Verlag, 220–41.

Evans DV (1981), 'Power water waves', *Annual Review of Fluid Mechanics*, **13**, 157–87.

EWG (2011), Edinburgh Wave Group Website. Available from: http://www.mech.ed.ac.uk/research/wavepower/ [Accessed February 2011].

Falcão A (2006), 'The history of and progress in wave energy conversion devices', *Proc. 9th World Renewable Energy Conference*, Florence, Italy.

Falnes J (2002), *Ocean Waves and Oscillating Systems*, Cambridge University Press.

Folley M and Cruz J, (2008), 'Power Take Off-Systems (Alternative Applications: Desalination)', in Cruz J, *Ocean Wave Energy – Current Status and Future Perspectives*, Springer-Verlag, 261–78.

Gieras JF (2008), *Advancements in Electrical Machines*, Power Systems, Springer.

Graw K (1995), *Wellenenergie – eine hydromechanische Analyse* (in German), Berlin, IGAW, BUGH Wuppertal.

Kelly HP (2000), 'The direct energy conversion method – the use of linear generators to capture and convert sea energy directly to electrical power', in *Wave Power – Moving Towards Commercial Viability*, IMechE Seminar Publication, UK.

Leishman JM and Scobie G (1976), 'The Development of Wave Power – A Techno-Economic Study', *NREL*, Department of Industry, UK.

Maki N, Takao T, Fuchino S, Hiwasa H, Hirakawa M, Okumura K, Asada M and Takahashi R (2005), 'Study of practical applications of HTS synchronous machines', *IEEE Transactions on Applied Superconductivity*, **15**, No. 2, 2166–9.

Miller C (2003), 'Wave Motors and Tide Machines at the Cliff House: 1886 to present'. Available from: http://www.outsidelands.org/wave-motor.php [Accessed February 2011].

Miller C (2004), 'A Brief History of Wave and Tidal Energy Experiments in San Francisco and Santa Cruz'. Available from: http://www.outsidelands.org/wave-tidal.php [Accessed February 2011].

Mueller MA, Baker NJ and Spooner E (2000), 'Electrical aspects of direct drive wave energy converters', *4th European Wave Energy Conference*, Aalborg University, Denmark.

Mueller MA (2002), 'Electrical generators for direct drive wave energy converters', *IEE Proc. Gener. Transm. Distrib.*, **149**, No. 4, 446–56.

Mueller MA and Baker NJ (2005), 'Direct drive electrical power take-off for offshore marine energy converters', *Proc. IMechE, Part: A Journal of Power and Energy*, **219**, 223–34.

Oregon (2011), Oregon State University – Division of Electricity Website. Available from: http://oregonstate.edu [Accessed February 2011].

Pelamis (2011), Pelamis Wave Power Website. Available from: http://www.pelamis-wave.com [Accessed February 2011].

Polinder H, Mecrow BC, Jack AG, Dickinson PG and Mueller MA (2005), 'Conventional and TFPM linear generators for direct-drive wave energy conversion', *IEEE Transactions on Energy Conversion*, **20**, No.2, 260–67.

Polinder H, Mueller MA, Scuotto M and Prado M (2007), 'Linear generator systems for wave energy conversion', *Proceedings 7th European Wave and Tidal Energy Conference*, Porto, Portugal.

Prado M (2008), 'Archimedes wave swing (AWS)', in Cruz J, *Ocean Wave Energy – Current Status and Future Perspectives*, Springer-Verlag, 350–61.

Prahlad H, Kornbluh R, Pelrine R, Stanford S, Eckerle J and Oh S (2005), 'Polymer power: dielectric elastomers and their applications in distributed actuation and power generation', *Proc. International Conference on Smart Materials Structures and Systems*, India.

Ruellan M (2007), *Méthodologie de dimensionnement d'un système de récupération de l'énergie des vagues* (in French), PhD thesis, École Normale Supérieure de Cachan, France.

Salter SH, Taylor JRM and Caldwell NJ (2002), 'Power conversion mechanisms for wave energy', *Proc. IMechE Part M, Journal for the Maritime Environment*, **216**, 1–27.

SARA (2011), Scientific Applications & Research Associates, Inc.'s Website. Available from: http://www.sara.com/RAE/ocean_wave.html [Accessed February 2011].

Scheijgrond PC, Schaap A, Sustronk BJ, Versteegh JR and Rossen EA (2006), 'Performance prediction of a combined Wells-Darrieus rotor with model tests and a computational vortex model', *Proc. World Maritime Technology Conference (WMTC)*, London, UK.

Taylor J (2008), 'Power Take-Off Systems (Hydraulics)', in Cruz J, *Ocean Wave Energy – Current Status and Future Perspectives*, Springer-Verlag, 241–60.

Tedd J, Friis-Madsen E, Kofoed JP and Knapp W (2008), 'Wave Dragon', in Cruz J, *Ocean Wave Energy – Current Status and Future Perspectives*, Springer-Verlag, 321–35.

Uppsala (2011), Uppsala University – Division of Electricity Website. Available from: http://www.el.angstrom.uu.se/ [Accessed February 2011].

WAVEC (2011), Wave Energy Center's Website. Available from: http://www.wavec.org/index.php/17/technologies/ [Accessed February 2011].

Wavedragon (2011), Wavedragon's Website. Available from: http://www.wavedragon.net [Accessed February 2011].

Wavestar (2011), Wavestar's Website. Available from: http://wavestarenergy.com [Accessed February 2011].

Yano M, Abe S and Ohno E (2004), 'History of power electronics for motor drives in Japan', *Proc. IEEE Conference on the History of Electronics*, UK.

9

Case study of the Archimedes Wave Swing (AWS) direct drive wave energy pilot plant

M. PRADO and H. POLINDER,
Delft University of Technology, The Netherlands

DOI: 10.1533/9780857097491.2.195

Abstract: This chapter provides a case study of the Archimedes Wave Swing (AWS) full-scale direct drive wave energy system prototype tested offshore Portugal in 2004. The AWS wave energy conversion concept is described, as well as the pilot plant design process, construction, installation and some experimental results from offshore tests.

Key words: linear generator systems, permanent magnet generators, ocean wave energy, direct drive, AWS, full-scale prototype, offshore tests.

9.1 Introduction

This chapter addresses the application of direct drive machines in wave energy conversion through a case study of the Archimedes Wave Swing (AWS) full-scale prototype. The AWS wave energy conversion concept and its full-scale prototype are described. There is then a more detailed discussion of the linear generator of the AWS prototype, looking at design, construction and installation aspects. The chapter ends with a resume of the results from offshore tests of the AWS pilot plant.

9.2 AWS wave energy converter

The AWS was the first multi-MW direct drive wave energy converter built and tested offshore. An overview of its concept development will be given, as well as the full-scale pilot plant design, construction and offshore tests.

9.2.1 The concept

The idea behind the AWS goes back to 1993, when F. Gardner and H. van Breugel started to work on the development of wave energy conversion technology (Rademakers *et al.*, 1998). The principles guiding the development were:

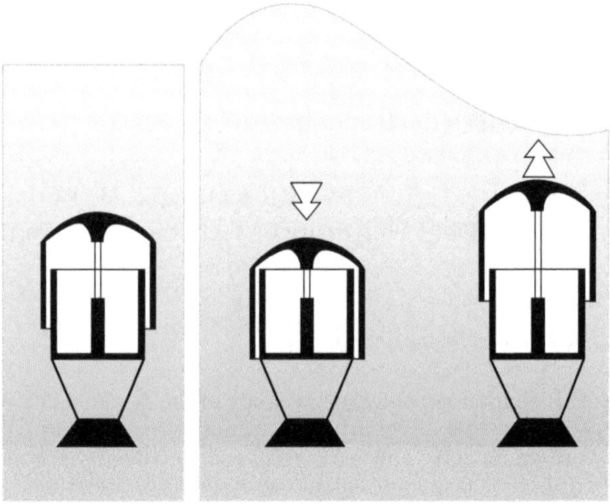

9.1 The AWS operating principle.

1. Submerged device to better survive storms.
2. An adjustable natural period of the captor to enhance energy absorption.
3. Simplicity of the design (low component system) to reduce failure risk.
4. Power levels above 1 MW.

After small-scale tests in 1997 at Delft, the Netherlands (scale 1:20) and in 1998 at Cork, Ireland (scale 1:50) (Rademakers *et al.*, 1998), the AWS concept started to mature into the device illustrated in Fig. 9.1.

The captor consists of a submerged piston, whose vertical oscillating motion is excited by the waves passing over it. To reduce the captor's reactance and achieve an adjustable natural period of the captor's dynamics, a gas spring is added inside the chamber of the piston. The stiffness of the gas spring can be adjusted by changing its volume through pumping water into or out of the piston chamber.

The ratio of the piston's stroke to its diameter is as high as possible in order to reduce the mechanical impedance of the captor, which favours the application of electrical power take-offs (PTOs). Different mechanisms to transform the linear motion into rotational were studied, so that a standard rotating electrical generator could be used as a PTO (Hulshof, 1999). Due to the complexity of these solutions, a direct drive PTO consisting of a linear electrical generator controlled by an AC/AC converter was chosen as the most appropriate option. Water brakes were added as a safety measure to damp the motion of the captor and keep it safely within its stroke, even when there is a PTO or grid failure. Other technical aspects of the technology, such as the mooring system and the sealing of the piston chamber, were still open issues.

9.2.2 Pilot plant

Although the small-scale models, backed up by mathematical models (Rademakers *et al.*, 1998), showed the technical and economic feasibility of the AWS concept, it was important for the development of the technology to demonstrate it and gain offshore experience at full scale. The idea was to build a pilot plant that was easy to deploy, test it under different sea conditions over a period of several months and then dismantle it. Important knowledge would be gained from:

1. the pilot plant construction, installation and operation;
2. cross-checking the models used in the design with the measured data;
3. wearing/corrosion processes after analysis of the dismantled parts.

In 2000, with finance in place, the AWS pilot plant design process began. The design was coordinated by Teamwork Technology BV and involved several companies (Alstom, W.E. B.V), research institutes (ECN) and universities (TUDelft – the Netherlands, and IST – Portugal). The overall design specifications were:

- Captor diameter: 9.5 m.
- Captor stroke: 7 m.
- Maximum linear generator force: 1 MN.
- Maximum linear generator power: 2 MW.
- Maximum speed: 2.2 m/s.
- Braking system: water brakes with continuous adjustable damping and low impact fail-safe end-stop braking.
- Test location: approximately 6 km offshore North Portugal, water depth 41–4 m.
- Deployment: easy and reversible (it should be possible to submerge or emerge the device in 1 day).
- Test period: less than 1 year.
- Licensing constraints: after the trials, no traces of the device are allowed at the test location.

Aspects of the technology not yet finalised were temporarily resolved with low-tech solutions that would not require much further design and development effort, reducing failure risk during the tests:

- **Piston chamber sealing** was achieved by having a sufficiently high internal air pressure to keep the water out of the device, eliminating the need for rubber seals.

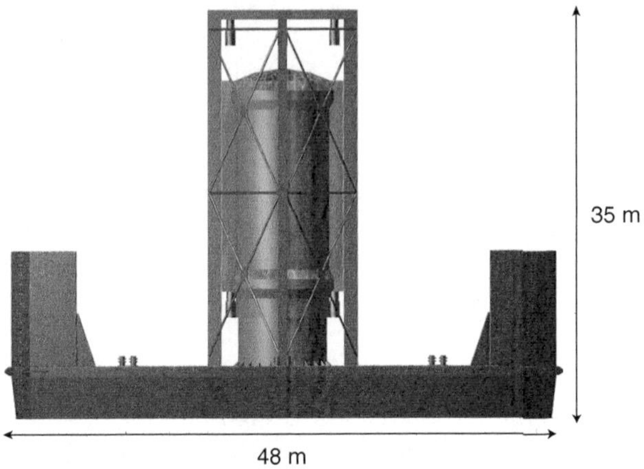

9.2 AWS pilot plant synthetic image: moving piston with guiding structure and pontoon (longitudinal side view).

- **Mooring** consisted of a submersible pontoon with four stabilizing towers and ballasting chambers that would allow easy mob/demob operations and avoid further environmental impacts on the seabed. The motion of the moving piston would be constrained by an outer guiding structure directly connected to the pontoon.

The final overall dimensions of the full-scale pilot plant including the pontoon were: length 48 m, width 28 m and height 35 m (see Fig. 9.2).

The AWS pilot plant was connected through a 6 km-long submarine cable to a land station on shore, where the controls and AC/AC converter were located. The submarine cable not only transmitted the generated power to shore but also supplied power from the shore to all auxiliary systems installed inside the device (instrumentation, pumps, valves, etc.) and provided fibre-optic communication between the land station and the device.

At the land station, in addition to the AC/AC converter, there was a capacitor bank to improve the power factor, a resistor bank that could take over the converter in case of grid loss or converter failure and a shunt to brake electrically the linear generator, when it was in stop mode. A simplified diagram of the overall system tested in Portugal is shown in Fig. 9.3.

The system components were made in different countries: hull (Romania), control system and water brakes (the Netherlands), submarine cable (Italy), linear generator (the Netherlands and France) and converter (Germany). All parts were then assembled together at Viana do Castelo shipyard, northern Portugal, near the deployment location (see Fig. 9.4).

Although the construction of the device and land station was completed by the end of 2001, financial constraints and technical difficulties with the

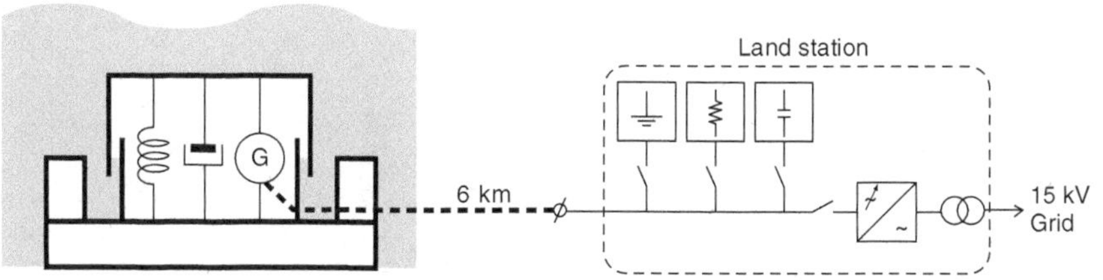

9.3 The AWS pilot plant connected to the land station.

9.4 The AWS pilot plant during final assembly at the Viana do Castelo Shipyard in Portugal, 2001.

submersion operation postponed the final submersion of the AWS pilot plant to May of 2004 (see Fig. 9.5). The details of the submersion operation are described by Prado and Gardner (2005).

Technical problems in communicating with the internal controls of the device further postponed the start of sea trials to mid-September 2004, when AWS was finally set to produce its first power (see Fig. 9.6). Furthermore, the water brakes were not fully operational due to lack of maintenance during the two and a half years that the pilot plant had to wait before being submerged. Therefore, for safety reasons, the water brakes were never switched off during the tests, which had a severe impact on the power performance of the device and also limited the range of possible results that could be extracted.

9.5 The AWS pilot plant at the beginning of the submersion operation, offshore North Portugal, May 2004.

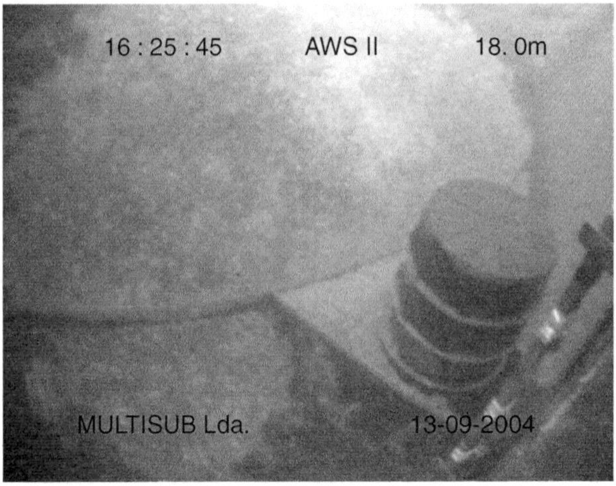

9.6 Underwater image of AWS's piston first motion, during the offshore tests in Portugal, September 2004.

The offshore test period turned out to be rather short. By the end of October 2004, the testing license issued by the Portuguese Maritime Authority expired and the tests had to be finished.

Despite all the constraints of the project, several tests were still performed under different conditions, enabling some fundamental features of the technology to be established. Some of the results obtained, in particular those concerning the linear generator, will be presented in this chapter, while a more detailed and complete analysis of the overall tests can be found in Polinder *et al.* (2005d) and Prado *et al.* (2005, 2006).

9.2.3 Further developments

After the 2004 offshore tests, the intellectual property in the AWS technology was transferred to the Scottish company AWS Ocean Energy Ltd, where it has since been further developed (Fitzgerald and Grey, 2008; AWSOCEAN, 2011).

9.3 AWS pilot plant power take-off (PTO): design and construction

The AWS pilot plant power take-off unit consisted of a linear generator grid connected through an AC/AC converter. This section will focus on the details of the generator design, its construction and installation.

9.3.1 Generator design

The main requirements for the generator system of the pilot plant of the AWS were listed in the previous section. During the design process, the following choices were made in order to match the requirements as closely as possible (Polinder *et al.*, 2004, 2005a, 2005b):

1. It is nearly impossible and extremely expensive to make the generator large enough to handle all possible forces generated by waves. Therefore, the AWS also has water dampers that can make very high forces. This implies that the generator can be designed as a compromise between energy yield and cost.
2. The linear generator that converts the mechanical energy into electrical energy is a permanent magnet (PM) generator because this generator type has a rather high force density and a rather high efficiency at low speeds (Polinder *et al.*, 2005b).
3. The magnets are on the translator that moves up and down, so that there is no electrical contact between the moving part (the translator) and the stator, which is important because such an electrical connection suffers from heavy wear.
4. The generator is flat. Although round generator constructions are a better fit in the construction of the AWS (Joseph and Cronje, 2008), for a single generator for a pilot plant, it was much cheaper to remain close to existing production technology and build a flat generator.

5. There is only one slot per pole per phase. Increasing this number leads to larger pole pitches, resulting in thicker yokes and a higher risk of demagnetization. Decreasing this number (using fractional pitch windings) leads to additional eddy-current losses due to additional space harmonics.
6. The magnets are skewed to reduce cogging.
7. The translator with the magnets is only a few metres longer than the stator in order to reduce cost. This means that in the central position, the magnets of the translator completely overlap the stator so that the maximum force can be exerted, but at the extremities, the magnets only partly overlap the stator, as depicted in Fig. 9.7.
8. To balance the attractive forces between stator and translator, the generator is double-sided, as depicted in Fig. 9.7.
9. For cooling the stator of the generator, a water cooling system was implemented.
10. The power electronic converter for the grid connection is placed on shore so that problems with the power electronics and the control can easily be solved. A 6 km-long cable connects the generator terminals to the converter on shore.
11. An onshore current source inverter is used for the utility grid connection. A voltage source inverter would have the advantages of better control characteristics, better power factor, better generator efficiency and higher forces (and energy yield) (Polinder *et al.*, 2004). However, the current source inverter appeared to be an easier and cheaper option.
12. Coatings are used to protect the generator against the aggressive environment.

The generator design was optimized from these starting points using analytical design equations (Polinder *et al.*, 2004, 2005a, 2005b). In the optimization, generator cost had to be minimized, while the energy yield had to be maximized, so both were included in the optimization criteria. The use of analytical design methods was possible because the flux density level in the generator was low enough to avoid significant saturation. Although further optimization of the generator using finite element methods was possible, this was not done due to time constraints.

Generator force and mechanical impedance

If the cogging and iron losses are ignored, the force of the generator can be obtained by dividing the total electrical power delivered by the generator to each of its phases by the speed of generator. When the speed of the generator is constant, the induced voltage will have a constant frequency proportional to the speed, that is

$$\omega_E = \frac{\pi}{\tau_P} v,$$

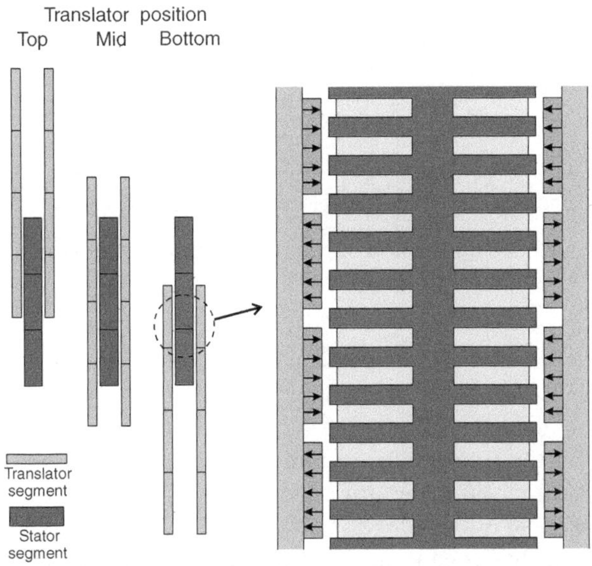

9.7 Right: Overlapping of translator and stator for different positions. Left: Section of four pole pitches of the linear PM generator. Middle: The stator with stator iron and coils in the stator slots in between. The left and right parts are the translators with the magnets with arrows indicating the magnetization direction.

where τ_P is the pole pitch. From the steady-state analysis of the electrical circuit of the generator, and assuming a balanced three-phase load, it is possible to derive the following analytical expression for the force–speed relation of the generator:

$$F_{PTO} = \frac{3}{2}\left(\hat{\lambda}_{pm}\frac{\pi}{\tau_P}\right)^2 \cdot \frac{\cos\left(\angle Z_E\left(\omega_E\right)\right)}{\left|Z_E\left(\omega_E\right)\right|}v,$$

where

- $\hat{\lambda}_{pm}$ is the amplitude of the first harmonic of the flux linkage due to the permanent magnets,
- Z_E is the total electrical impedance of the electrical circuit of one phase of the generator.

The force is not necessarily proportional to the speed, since in general the electrical impedance (both magnitude and phase) may change with frequency. If the electrical impedance is purely resistive, the relation between the generator's force and speed will become linear. The presence of a net

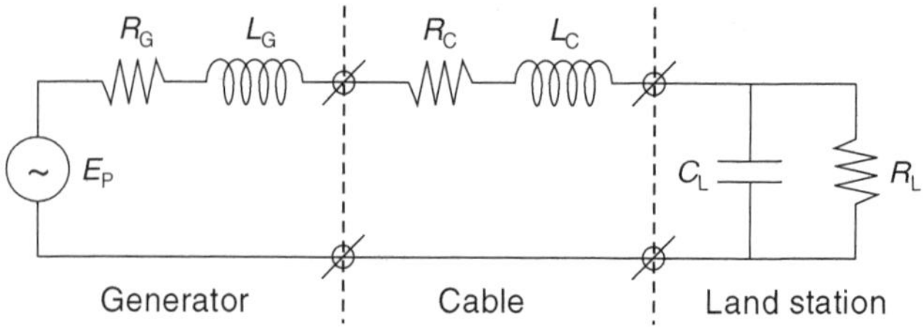

9.8 Equivalent single-phase circuit model of the linear generator, cable and landstation, with a resistive end load.

reactance in the electrical circuit will introduce a non-linearity into the force–speed relation. In general, the higher the net reactance of the electrical circuit, the smaller the generated force, since it increases both the modulus and angle of the total electrical impedance. This is why power factor correction of the internal generator's inductance is important to enhance the force–speed curve of the generator.

In the case of the linear generator tested in Portugal, the equivalent electrical circuit of the stator is depicted in Fig. 9.8.

This circuit includes the internal resistance and inductance of the generator, the resistance and inductance of the 6 km power cable, the power factor correcting capacitor and the load resistance connected to the end of the cable. The total electrical impedance of the generator circuit is thus given by

$$Z_E(\omega_e) = (R_G + R_C) + j\omega_e(L_G + L_C) + \frac{R_L}{1 + j\omega_e C_L R_L}$$

which is shown in Fig. 9.9 for different values of load resistance.

Up to frequencies of 1 Hz (i.e. speeds up to 0.2 m/s), the electrical impedance is reasonably resistive. However the translator's speed can be as high as 2.2 m/s, which means the electrical frequency can go up to 11 Hz, where significant phase and amplitude variations of the electrical impedance can be observed. For a load resistance close to 6Ω, the electrical impedance is reasonably constant up to 11 Hz, so an approximately linear relation between force and speed is expected.

The force–speed relation of the linear generator at constant speed with the electrical impedance described above is shown in Fig. 9.10.

As the speed increases and the load resistance deviates 6Ω, the linear force–speed relation becomes more distorted. For smaller load resistances, the inductance of the generator dominates the net reactance of the electrical circuit, which causes the force to saturate and decrease as speed increases. For higher load resistances, the power factor correction effect of

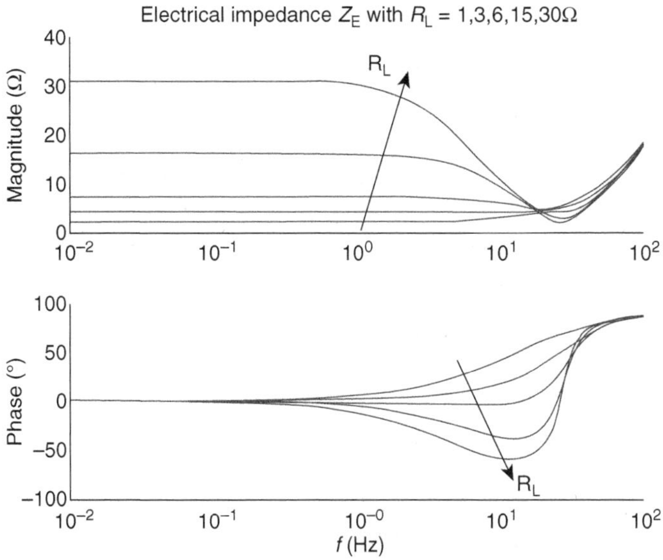

9.9 Bode diagram of the total electrical impedance of the equivalent single-phase circuit model of the generator, for different load resistances.

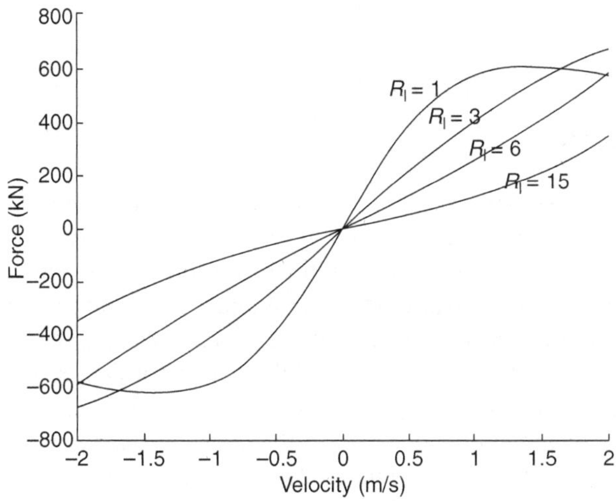

9.10 Force–speed characteristic of the linear generator for constant speed and different resistance loads (with total overlap between translator and stator).

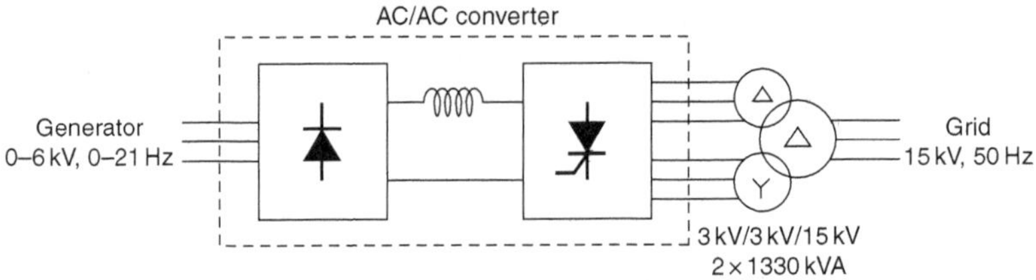

9.11 AC/AC converter used to grid connect the linear generator.

the capacitor becomes visible, which results in a further increase of the force as speed increases. This boost of the generator force is, however, limited up to the point when the inductance of the generator starts dominating the net reactance of the electrical circuit as speed further increases.

For low velocities around the mid-position (total overlap), the force is reasonably linear with speed and therefore the damping coefficient of the PTO is given by (Prado *et al.*, 2006)

$$\beta_{\text{PTO}}(R_L) = \left.\frac{\partial F_{\text{PTO}}(v, x, R_L)}{\partial v}\right|_{v=x=0},$$
$$\approx \frac{1.64 \times 10^6 \text{ Nsm}^{-1}\Omega}{R_G + R_C + R_L}$$

with $R_G + R_C \approx 1\Omega$.

9.3.2 Converter

The converter consisted of a six-pulse input diode rectifier bridge, an intermediate DC-link choke and a 12-pulse output thyristor bridge (see Fig. 9.11).

With this converter topology, energy can only flow from the generator to the grid but not vice versa, that is, the linear generator can only work as a generator and not as a motor. Thus the mechanical impedance of the generator is restricted to the real axis, and the generator can only work as a damper. However the converter allows the damping coefficient of the linear generator to be changed. For this reason the converter was equipped with an internal control loop that automatically adjusted the firing angles of the thyristor bridge in order to keep the DC-link resistance R_{DC} as constant as possible at the desired resistance setpoint. Despite the intrinsic limitations of the controller in keeping the DC-link resistance constant and the nonlinearities of the rectifier bridge, the converter can be approximated from the electrical point of view as an adjustable linear resistive load R_L directly connected to the generator (see Fig. 9.12).

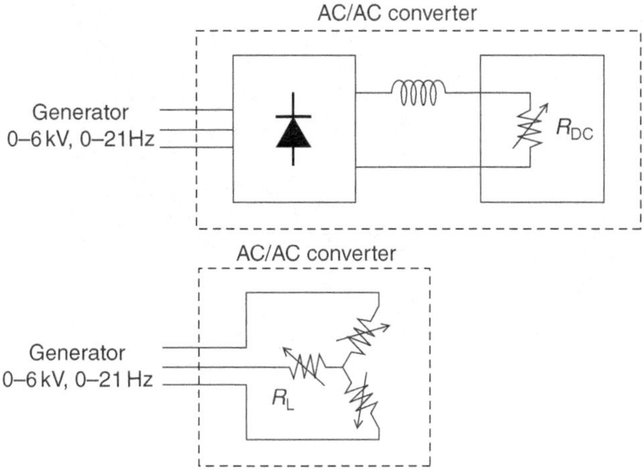

9.12 AC/AC converter equivalent models: thyristor bridge with constant input resistance (top) and pure resistive load (bottom).

Neglecting the losses in the rectifier and DC coil and assuming a balanced three-phase system, both circuits dissipate the same average power if

$$R_L = \left(1 + \frac{3\sqrt{3}}{2\pi}\right)^{-1} R_{DC} \approx 0.55 \times R_{DC}.$$

9.3.3 Construction and installation

For practical production reasons, the stator and translator of the linear generator were each split into two equal parts. Each stator part consisted of three segments, and each translator part consisted of four segments (see Fig. 9.7).

The translators and stators were built in different places because of their very different characteristics. The translators were built in the Netherlands by Bakker Magnetics BV. Figure 9.13 depicts a part of a translator segment about 2 m long and 1 m wide. The stator was built by Alstom in Nancy, France (see Fig. 9.14).

Translator and stator segments were transported to Portugal, where they where assembled together inside the pilot plant at the Viana do Castelo Harbour (see Fig. 9.15).

Figure 9.16 depicts the converter, together with the resistor banks, the power factor correction capacitors and the power transformers installed at the land station in Portugal.

9.13 Translator segment with magnets in production in the Netherlands.

9.14 Stator segment with coils at Alstom's factory in France.

9.4 AWS pilot plant power take-off (PTO): test results

Several tests were performed on the linear generator during different phases of the project, and some of these are described below.

9.4.1 Onshore: assembly

During assembly of the linear generator, while the translator was being lowered along the stator, the voltages (electromotive forces) induced by the translator in the stator were measured. Figure 9.17 shows the measured no-load voltage of the generator at a very low and not exactly constant speed (a few centimetres per second).

9.15 Linear generator's stator (top) and translator segments (bottom) being assembled together inside the AWS pilot plant.

9.16 Converter with resistor banks, power factor correction capacitors and power transformers installed at the land station in Portugal.

Before t = 40 s, the overlap between stator and translator is constant. Around t = 45 s, there is a stop for a few seconds. After t = 50 s, the overlap between stator and translator reduces, resulting in a decreasing no-load voltage while the frequency and speed remain approx. constant. The flux

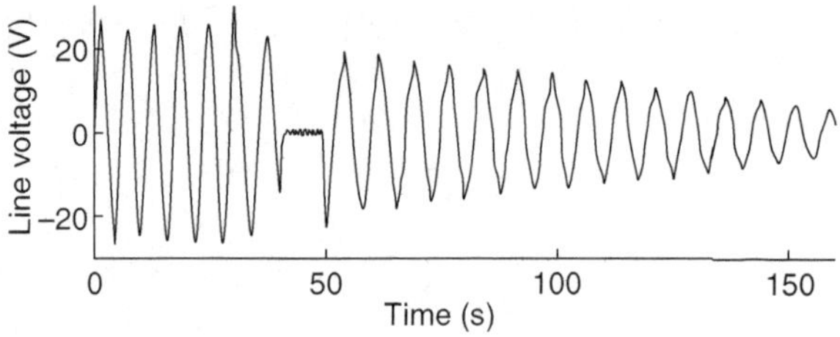

9.17 Measured no-load voltage of one of the generator parts, while a translator segment was being lowered along it, during assembly of the linear generator at Viana do Castelo Shipyard, Portugal, September 2001.

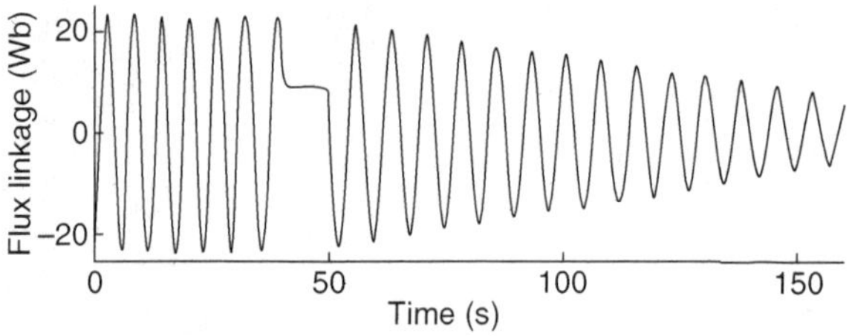

9.18 Calculated flux linkage from measured no-load generator output voltage (September 2001).

linkage was obtained by integrating the no-load voltage (see Fig. 9.18). This flux linkage appears to be reasonable sinusoidal, which is mainly due to the magnet skewing. The measured flux linkage (for full stator–translator overlap) was approx. 23.5 Wb, which was within 3% of the design specifications (Polinder *et al.*, 2004).

The resistance and leakage inductance of the generator were also determined by measuring the current response to step voltage signals applied to the stator (see Fig. 9.19).

System identification analysis of the equivalent RL model of the generator was carried out, based on the voltage and current signal measurements. The resistance and inductance values identified were approximately 0.29Ω and 31 mH respectively, which differ by less than 11% from the design specifications (Polinder *et al.*, 2005a).

9.4.2 Offshore: resistive load

On 2 October 2004, the generator was connected to the 6Ω resistor bank. The mechanical impedance imposed by the generator on the captor's dynamic was therefore constant. The total power dissipated by the resistor bank and the total power output of the generator were calculated by measuring the

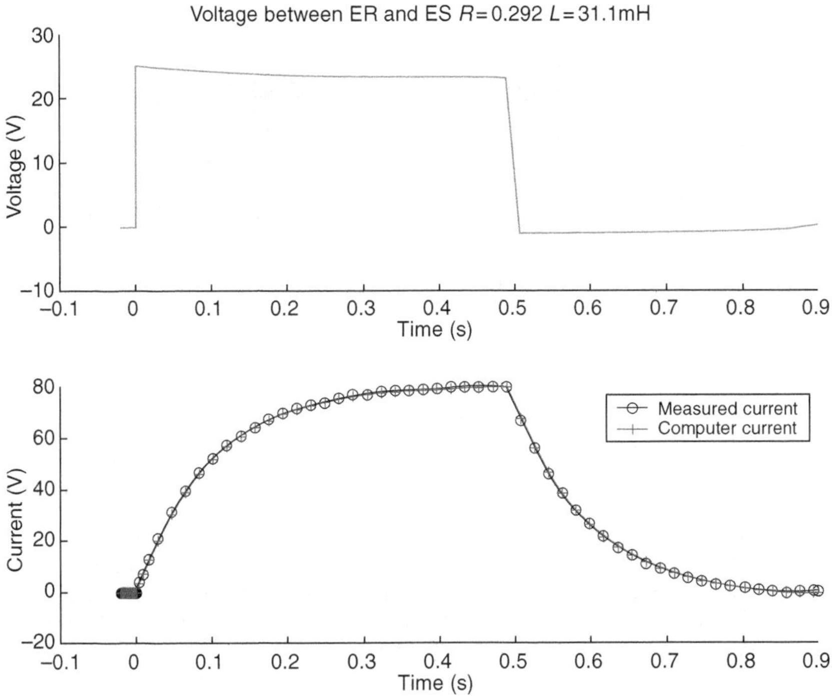

9.19 Measured voltage (top) and measured/computed current (bottom) from step voltage tests on the generator (September 2001).

currents at each phase. Figure 9.20 shows the currents and the corresponding power levels for a complete oscillation of the captor of approximately 10 s (1 wave period).

From Fig. 9.20, the following observations can be made:

1. For the complete wave cycle, there are two generated power pulses (i.e. the AC component of the power signal has twice the frequency of the wave).
2. The power output of the generator is higher than the power dissipated at the resistor bank, due to the resistive losses along the 6 km underwater cable.
3. Since both the frequency and amplitude of the induced voltage in the stator are proportional to the speed of the generator, it is also clear, from the measured current signals, that the power levels increase with the generator's speed. Power peaks occur therefore whenever the generator's speed reaches its maximum and the output power level is zero whenever the position of the generator reaches its extreme position (i.e. speed is zero).

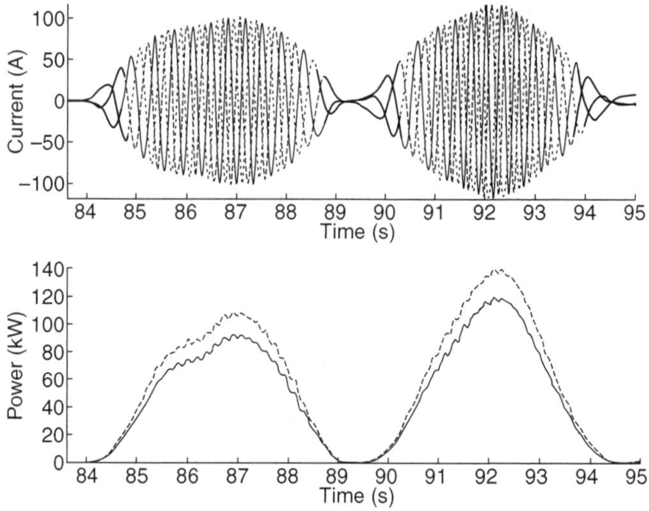

9.20 Test results with generator connected to the resistor bank (2nd of October 2004). Top graphic: resistor current measurements (three phases). Bottom graphic: Solid line corresponds to dissipated power at resistor bank and dashed line corresponds to the power output of the generator.

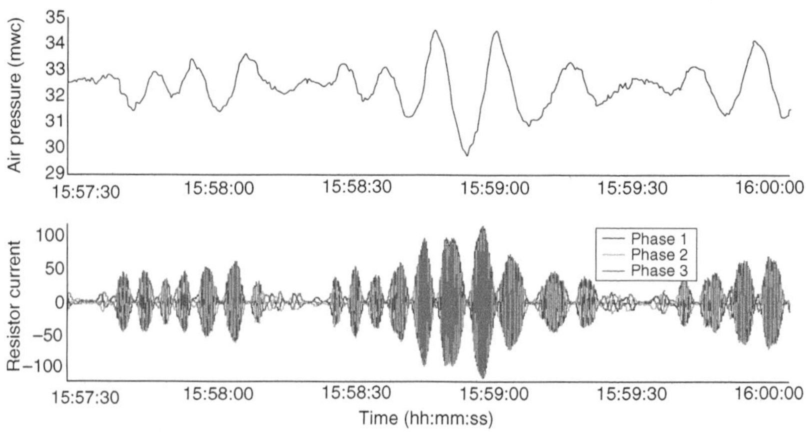

9.21 Test results with the generator connected to the resistor bank (2 October 2004). Top graphic: measured air pressure inside of the AWS piston chamber. Bottom graphic: measured electrical currents at the resistor bank (three phases).

This direct link between the motion of the device and the electrical signals of the generator can be better visualized in Fig. 9.21, which shows the measured air pressure inside the AWS's piston chamber and the measured electrical currents at the resistor bank.

Bearing in mind that the air pressure signal is related directly to the position of the generator (the higher the position, the smaller the pressure inside the piston chamber), the following observations can be made from Fig. 9.21:

1. The maxima of the current-signals envelope coincide with the extremes of the air pressure derivative, that is, maximum generator speed.
2. The minima of the current-signals envelope coincide with the extremes of the air pressure, that is, zero generator speed.
3. The current-signals envelope has twice the frequency of the air pressure signal, since the former is directly related to the generated power and the latter is directly related to the piston motion.

Since the induced-voltage signals are amplitude modulated by the generator's speed and phase modulated by the generator's position, it is possible to demodulate the current signals and recover the generator motion (position or speed) from them. This would allow the implementation of a sensorless motion control of the linear generator, if the estimate of the motion is sufficiently accurate. One way of determining the position of the generator from the measured currents is to use the flux linkage computation for each of the phases from the corresponding current signals (Polinder *et al.*, 2005d). Since no direct measurement of the generator position was available during the tests, one way of cross-checking the flux-based position estimate is to compare it to another estimated position based on a different measurement. For this purpose, the air pressure measurement shown in Fig. 9.21 can be used, since the position of the generator is related to the air pressure inside the piston chamber through a non-linear adiabatic air-spring curve (Prado *et al.*, 2006).

The generator position signals derived from the current and air pressure measurements using these estimation methods are shown in Fig. 9.22.

As can be seen from Fig. 9.22, both estimates show a reasonably good fit, although some deviations up to 20 cm are observed between the signals. This can be due to various factors such as noise in the measurements, signal filtering and errors in the models used for the estimation. Nevertheless the resulting data coherence is quite acceptable, considering that the estimates are based on measurements of completely different natures (air pressure and electrical currents).

9.4.3 Offshore: grid connected

On 3 October 2004, the generator was connected to the grid via an AC/AC converter. The input currents of the converter were measured and are shown for just half of one wave cycle in Fig. 9.23.

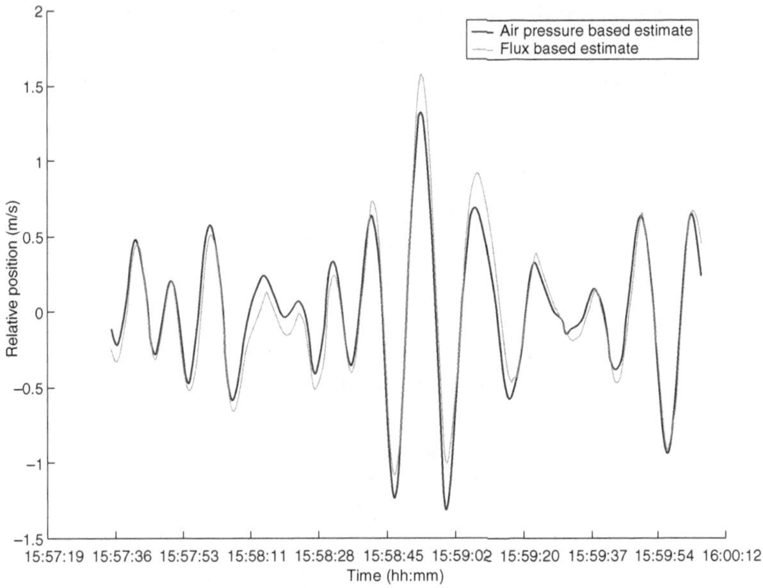

9.22 Generator position estimates based on air pressure measurements inside the AWS piston chamber and based on electrical current measurements at the resistor bank.

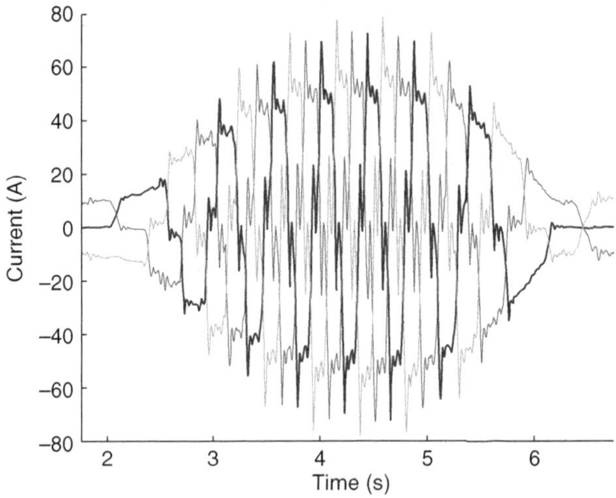

9.23 Measured currents at the input of converter during tests with the generator connected to the grid through the converter for one half wave cycle (3 October 2004).

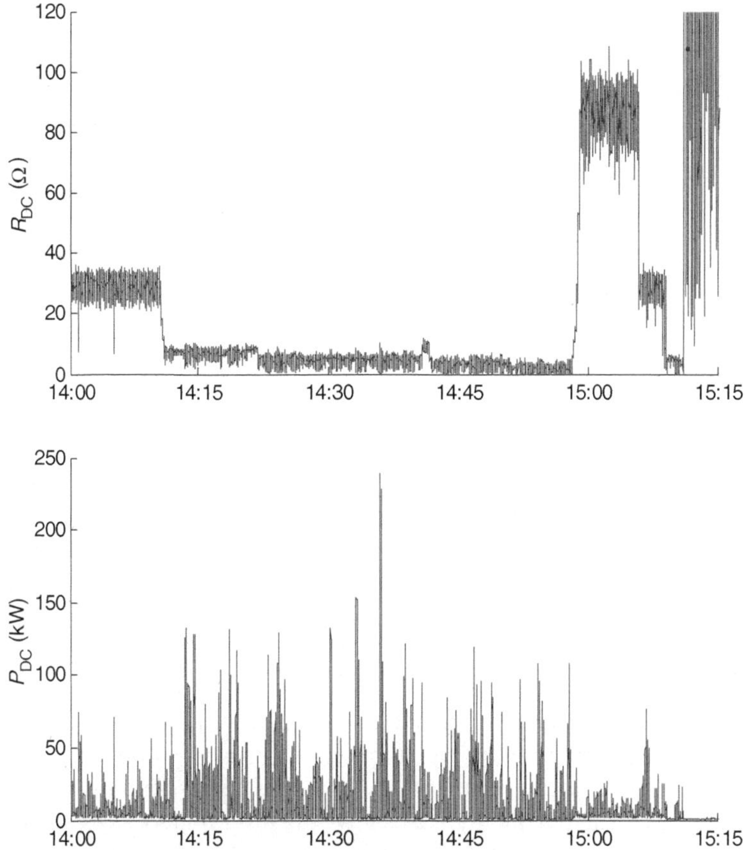

9.24 Test results with the generator connected to the grid via the converter (3 October 2004).

As when the generator was connected to the resistor bank, the frequency and amplitude of the measured currents increase and decrease during each half-cycle, but now in a rather 'noisier' way. This is a direct result of the non-linear behaviour of the input diode bridge of the converter, which causes the input converter current to switch on and off periodically at each phase. The ripples observed correspond to the electrical transient at each switching step, due to the resonance of existing capacitors (power factor correction) and inductances (DC link, cable and generator) in the electrical circuit.

One big advantage of using the converter was the possibility of changing the mechanical impedance of the linear generator by adjusting the DC-link resistance of the converter. Different setpoints for the DC-link resistance R_{DC} were given to the converter. The smaller the resistance, the higher the

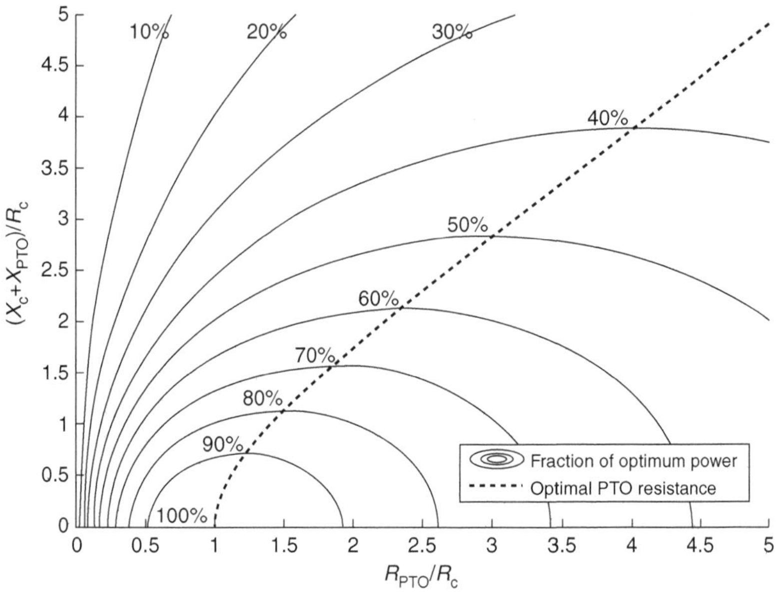

9.25 Test results with the generator connected to the grid via the converter (3 October 2004).

mechanical impedance of the PTO applied to the captor dynamics. The voltage and the current at the DC link were measured, and from these the total power flowing at the DC link P_{DC} and the corresponding resistance R_{DC} could be calculated. Figure 9.24 shows the instantaneous resistance and the corresponding power at the DC link of the converter for a period of approximately 1 h 15 min (≈450 wave periods), during which there was no significant change in sea conditions.

The irregularity of the waves can be seen clearly from the power signal depicted in Fig. 9.24 (notice that for each wave cycle there are two power pulses). The average power was roughly 24 kW, while the peak power was approximately 240 kW. Notice that these power levels are small compared to the 2 MW nominal power of the device, since the water brakes were always on during the tests. For the prevailing sea conditions on 3 October, average power levels were expected to be 200 kW if water brakes were to be switched off (Prado *et al.*, 2005).

Furthermore, Fig. 9.24 shows that for higher values of R_{DC} lower power outputs are observed at the converter. This is a direct consequence of the fact that mechanical impedance of the PTO decreases when R_{DC} increases. Since the captor impedance was higher than the PTO impedance (water brakes were on for safety reasons), the power output of the PTO decreases when its mechanical impedance decreases, as shown in Fig. 9.25 and also discussed in page 181 in Chapter 8.

The results of these tests were used to validate important parts of the hydromechanical and electrical models. However, full validation of all aspects was not possible due to the constraints of the water brakes and the limited test period (Prado *et al.*, 2006).

9.5 Conclusion

The experimental results from the offshore tests described above validate important aspects of the hydromechanical and electrical models of the AWS. However, inoperability of the water brakes meant that the prototype could not be tested over its full power range. Due to the limited time period of the offshore tests, it was not possible to acquire maintenance and reliability data relating to the linear generator.

9.6 Acknowledgement

This work was supported in part by a Marie Curie Initial Training Network of the European Community's Seventh Framework Program under contract number PITN-GA-2008-215414, the Wavetrain2 program.

9.7 References

AWSOCEAN (2011), AWSOCEAN's Website. Available from: http://www.aws-ocean.com/ [Accessed February 2011].

Fitzgerald J and Grey S (2008), 'Evolution of the AWS WaveswingTM MkII Concept', *Proc. 2nd International Conference on Ocean Energy (ICOE)*, Brest, France.

Hulshof G (1999), *Elektriciteit uit Golven* (in Dutch), Hogeschool Enschede, The Netherlands.

Joseph DM and Cronje WA (2008), 'Design of a double-sided tubular permanent-magnet linear synchronous generator for wave-energy conversion', *COMPEL: The International Journal for Computation and Mathematics in Electrical and Electronic Engineering*, **27**, 154–69.

Polinder H, Damen MEC and Gardner F (2004), 'Linear PM generator system for wave energy conversion in the AWS', *IEEE Transactions on Energy Conversion*, **19**, 583–9.

Polinder H, Damen MEC and Gardner F (2005a), 'Design, modelling and test results of the AWS PM linear generator', *European Transactions on Electrical Power*, **15**, 245–56.

Polinder H, Mecrow BC, Jack AG, Dickinson P and Mueller MA (2005b), 'Linear generators for directdrive wave energy conversion', *IEEE Transactions on Energy Conversion*, **20**, 260–7.

Polinder H, Mecrow BC, Jack AG, Dickinson PG and Mueller MA (2005c), 'Conventional and TFPM linear generators for direct-drive wave energy conversion', *IEEE Transactions on Energy Conversion*, **20**, No. 2, 260–67.

Polinder H, Damen MEC, Gardner F and Prado M (2005d), 'Archimedes wave swing linear permanent-magnet generator system performance', *Proc. 6th European Wave and Tidal Conference*, 383–7, Glasgow, Scotland.

Prado M and Gardner F (2005), 'Theoretical analysis of the AWS dynamics during submersion operation', *Proc. 6th European Wave and Tidal Conference*, 395–9, Glasgow, Scotland.

Prado M, Neumann F, Damen MEC and Gardner F (2005), 'AWS results of pilot plant testing 2004', *Proc. 6th European Wave and Tidal Conference*, 401–7, Glasgow, Scotland.

Prado M, Gardner F, Damen MEC and Polinder H (2006), 'Modelling and test results of the Archimedes Wave Swing', *Proc. IMechE, Part A: Journal for Power and Energy*, **220**, 855–68.

Rademakers LWMM, Van Schie RG, Schuitema R, Vriesema B and Gardner F (1998), 'Physical model testing for characterising the AWS', *Proc. 3rd Eur. Wave Energy Conference*, **1**, 192–9, Patras, Greece.

10
Application of high-temperature superconducting machines to direct drive renewable energy systems

O. KEYSAN, University of Edinburgh, UK

DOI: 10.1533/9780857097491.2.219

Abstract: After the discovery of high-temperature superconductors, superconducting machines became more feasible and have been applied to many applications such as ship propulsion systems. Superconducting machines are lighter, smaller and more efficient than conventional generators. In this chapter, mass of different direct drive generators systems have been compared. Direct drive superconducting generators have the potential to reduce the nacelle mass and the installation cost of a large offshore wind turbine. Ongoing commercial projects and the requirements for a reliable direct drive superconducting generator are both listed.

Key words: superconductivity, superconducting machines, offshore wind turbines, high-temperature superconductors.

10.1 Introduction

It is usually assumed that superconductivity is a cutting-edge technology. On the contrary, the phenomenon of superconductivity was first discovered over a century ago in 1911, by a Dutch scientist named Kamerlingh Onnes. Onnes was doing experiments on the physical properties of mercury at very low temperatures when he observed an abrupt decrease in its resistivity. Electrical resistances of conventional materials decrease as the material is cooled down. The resistivity converges to a non-zero value as the temperature is reduced down to absolute zero (–273°C). Superconductors differ from normal conductors as their electrical resistances plunge from a finite value to zero at a specific point which is called the critical temperature (T_c). The resistivity variation of a normal conductor and a superconductor with temperature is given in Fig. 10.1.

The first superconductor material discovered was mercury which has a critical temperature of 4.2 K. Since then many other superconductors have been discovered, such as lead and niobium. Superconductors are not exotic and

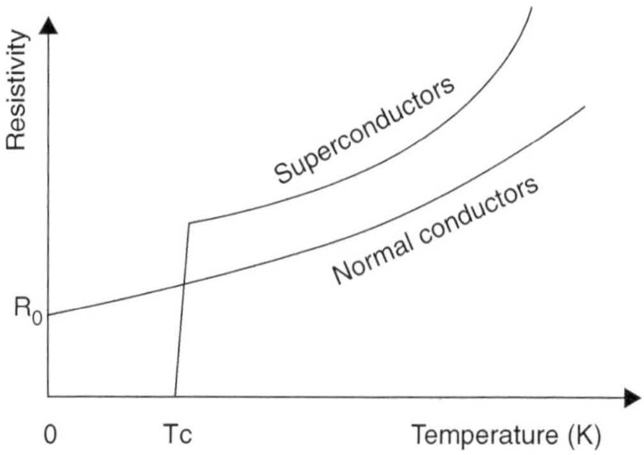

10.1 Variation of electrical resistivity with temperature for normal conductors and superconductors. T_c is the critical temperature of the superconductor.

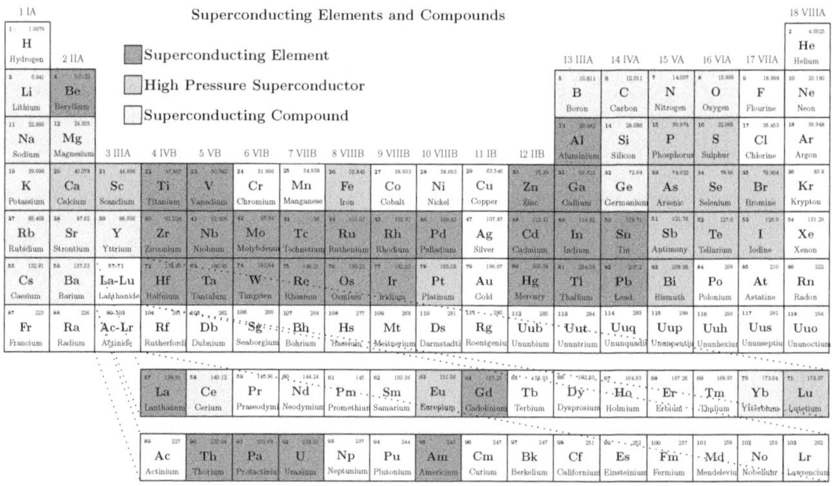

10.2 Known superconducting elements and elements that constitute superconducting compounds.

rare materials; in fact there are many superconducting elements and many more superconducting compounds. Some elements only become superconducting under high pressure. The superconducting elements and the elements that take part in superconducting compounds are shown in Fig. 10.2.

The most important feature of a superconductor is the critical temperature. There are many superconductors with a very low critical temperature, around a few Kelvins, which have no practical application. High T_c is desirable, but it is not the only attribute of a superconductor. Superconductivity is also limited by current density and magnetic field applied to the material. The critical temperature (T_c), the critical current density (J_c) and the critical magnetic

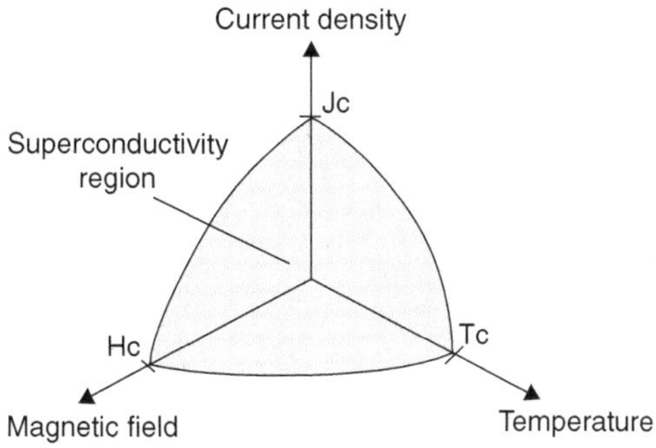

10.3 Boundaries of the superconductivity region (current density, temperature and magnetic field).

field strength (H_c) define the boundaries of the superconductivity state as in Fig. 10.3, and materials lose their superconductivity outside this region.

10.1.1 Properties of superconductors

Zero resistivity

The most distinctive property of superconductors is the zero resistivity. Unfortunately, this statement is only valid when a direct current (DC) is conducted. Superconductors dissipate power when conducting alternating current. This is due to the fact that superconductors have two different types of electrons: super-electrons and normal electrons (Kakani, 2009). As the temperature is reduced below the critical temperature (T_c), super-electrons begin to develop which are capable of conducting current without any resistance at all. The remaining electrons are called normal electrons and they suffer collisions and dissipate power as the current flows. The fraction of super-electrons increases as the temperature is further decreased and at 0 K all the conduction electrons become super-electrons (Kakani, 2009).

In a superconductor, current can be conducted by both normal and super-electrons. When a direct current is applied, the super-electron path short-circuits the normal electron path and no current is conducted through the normal electron path. On the other hand, for an alternating current, an electric field must exist to accelerate (and to decelerate) the electrons. The inertia of an electron is very small, but there is still a lag between current and electric field. This means the super-electrons introduce an inductive impedance. In other words, superconductors can be modelled by an ideal inductance in parallel with a resistance (Kakani, 2009) as given in Fig. 10.4. The inductive component creates a voltage potential under AC conditions and normal electrons conduct some of the current. As a consequence, the lossless transmission of superconductors is disrupted for alternating currents.

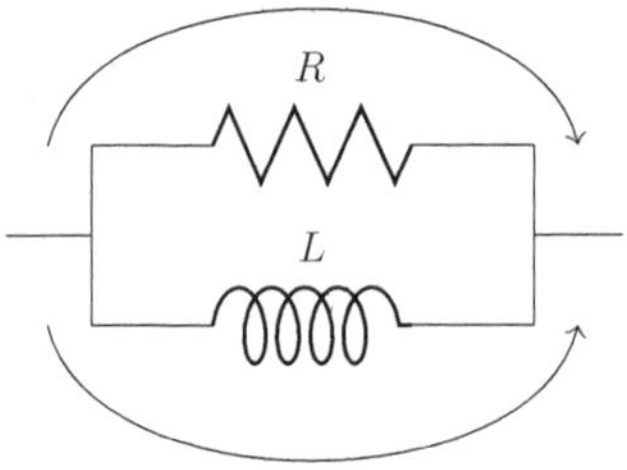

10.4 Electrical representation of a superconductor.

Perfect diamagnetism

A less-known property of superconductors is perfect diamagnetism. When a magnetic field is applied to a superconductor, non-decaying screening currents are induced on the surface of the superconducting material which creates an opposing magnetic field preventing flux from penetrating inside the superconductor. This behaviour, which is called the Meissner-Ochsenfeld Effect, was discovered in 1933. This effect can be expressed as:

$$B = \mu_0 (H + M) = 0 \tag{10.1}$$

where M is the magnetization. The equation implies perfect diamagnetism ($M/H = -1$) (Kakani, 2009). A special type of superconductor (so-called type II superconductors) allows some flux to penetrate inside the surface. This is called flux pinning and it gives stability in magnetic levitation. In other words, it is the Meissner effect that creates the levitation force and it is the flux pinning that gives stiffness and balance to the levitation. In Fig. 10.5 a superconductor levitating above a permanent magnet is shown.

Quench

Quench is the process of a superconductor transforming from superconducting state to normal state. It can be triggered by a change in temperature, current density or magnetic field. As a consequence, the superconductor goes from having no resistance to being resistive and it generates a large amount of joule heating which in return accelerates the quenching process. This positive feedback mechanism is one of the most critical issues and should be carefully addressed. Quench is usually a design challenge, but it is very useful for some applications. For example, resistive superconducting fault current limiters depend on this property to quickly limit the value of fault currents in a grid by instantly going from the superconducting state to the normal state.

10.5 A superconductor levitating above permanent magnets (Photo: Dr Kim Chan-Joong).

10.2 Common superconducting wire materials

There are many superconducting elements as presented in Fig. 10.2. However, only a fraction of these elements are used in electrical wire applications. For some materials the critical temperature is too low or the performance of the material is not satisfactory to be used as superconducting wire; for others the wire manufacturing process is too expensive. The most common materials that are used in superconducting wires can be listed as BSCCO, YBCO, MgB2, NbTi and Nb3Sn. Only the first three materials will be reviewed in this chapter.

10.2.1 Bismuth strontium calcium copper oxide (BSCCO)

Bismuth strontium calcium copper oxide (BSCCO, Bi_2 Sr_2 $CaCu_2$ O_8) is a copper oxide, which is a bad conductor at room temperature. The critical temperature for BSCCO is 108 K. BSCCO is the first high-temperature superconductor (HTS) to be used for making superconducting wires and it is the first HTS discovered that does not contain any rare-earth elements. BSCCO wires are usually called first-generation (1G) HTS wires. The comparison of a BSCCO tape with a similarly rated copper wire is presented in Fig. 10.6.

BSCCO itself has a ceramic structure and thus it is brittle and has low mechanical strength. In order to manufacture BSCCO in the form of tape, support from stronger metals is needed. This is achieved by embedding

10.6 Comparison of BSCCO tape with a similar rated copper wire (Courtesy of Bruker EST).

BSCCO filaments into silver matrix. The conductor is then rolled flat to form tapes. Although the superconducting material itself can be manufactured at a reasonable price, high silver content and multiple processing steps adds extra cost to BSCCO tapes (Schiferl *et al.*, 2006).

There are a few types of BSCCO compounds. The two most common types of BSCCO superconductors are: Bi-2212 ($Bi_2 Sr_2 CaCu_2 O_8$) and Bi-2223 ($Bi_2 Sr_2 Ca_2 Cu_3 O_6$). Bi-2212 has a critical temperature of 95 K which enables the use of liquid nitrogen as the coolant gas, but it is common to cool BSCCO wire down to 20–40 K to achieve higher current density. Bi-2223 has the highest critical temperature among BSCCO compounds (108 K). The usual tape is 4 mm wide and 0.2 mm thick and can carry 200 A at 77 K. Bi-2223 is a common superconductor wire for transformers, motors and generators.

10.2.2 Yttrium barium copper oxide (YBCO)

Yttrium barium copper oxide (YBCO, $YBa_2 Cu_3 O_7$) is also a copper oxide, like BSCCO. It was discovered in April 1986. YBCO, which has a critical temperature of 93 K, is the first superconductor with a critical temperature higher than the boiling point of nitrogen. This was a significant breakthrough because nitrogen is the most abundant element in the atmosphere and cheap to store. Moreover, liquid nitrogen-based cryocoolers are reliable and have been commonly used in industry for many years. The discovery of YBCO prompted an expansion of superconducting technology research.

Although YBCO was discovered before BSCCO, it took longer to develop a manufacturing technique for YBCO wires. YBCO is usually

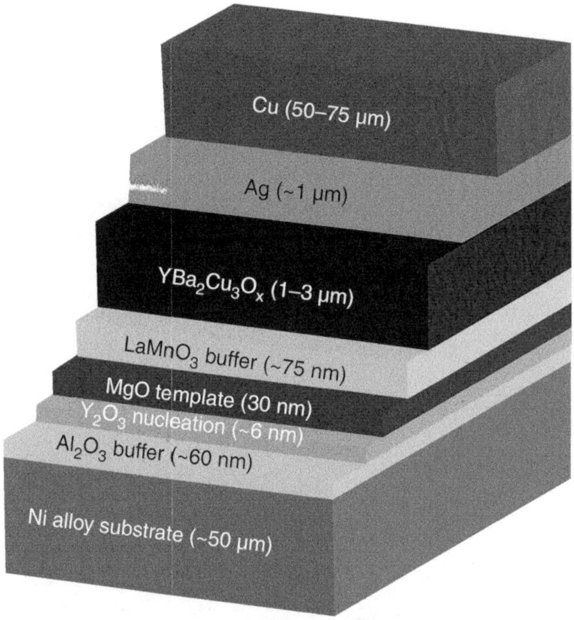

10.7 Layers in a YBCO-coated conductor (Gurevich, 2011).

referred to as second-generation (2G) superconducting wire. YBCO tapes are usually made up of multiple layers (see Fig. 10.7). They can be made in a continuous process, reel to reel, in a controlled chamber. The materials in the buffer, substrate and stabilizer layers are cheaper than silver. The combination of the lower raw material cost and simpler manufacturing methods makes 2G YBCO wires less expensive than 1G wires (Schiferl *et al.*, 2006). According to a study by Klaus *et al.* (2007), the cost of 2G HTS wires is expected to halve once the mass production techniques are established. This is supported by an industrial feasibility analysis by Converteam that predicts the cost of 2G HTS wires will be cheaper than current copper prices in a decade (Lewis and Muller, 2007). The same results are presented in Fig. 10.8 (SuperOx web page).

A major advantage of YBCO wire is its ability to remain at superconducting state with current densities (J_c) greater than 10^3 A/mm^2 in fields up to several Tesla. Also some manufacturing methods successfully applied on YBCO and YBCO-coated conductors have resulted in dramatic improvements in the quality and length of wire available, reaching lengths of several meters with high performance (Barnes *et al.*, 2005b). Kalsi (2011) states that the maximum available length of YBCO wire is 500 m.

The critical operation point of superconductors is non-linear and varies with temperature. The performance of the superconductor can be

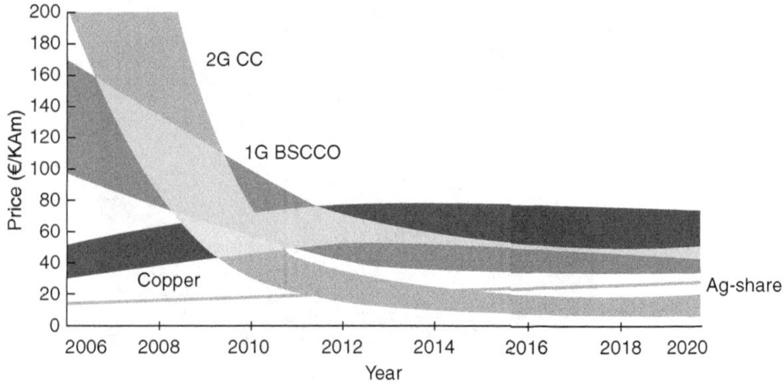

10.8 YBCO, BSCCO and copper wire price (€/kAm) projection (SuperOx, 2012).

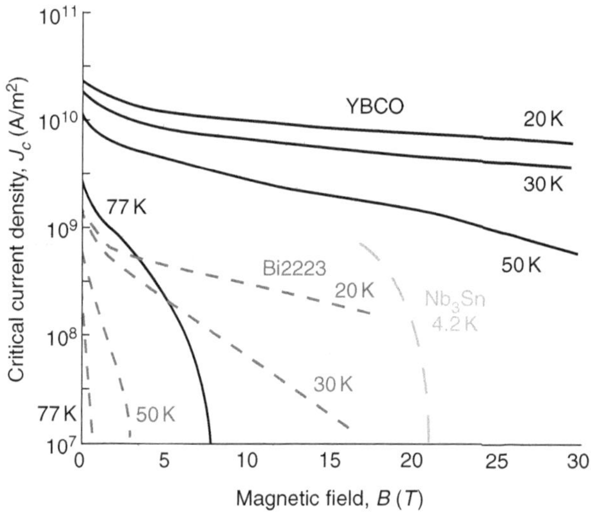

10.9 Performance comparison of some common superconductors (Ohsaki *et al.*, 2009).

significantly improved by cooling beyond the critical temperature. The characteristics of YBCO and BSCCO are given in Fig. 10.9.

In Table 10.1, the characteristics of BSCCO and YBCO are compared and they are evaluated for HTS generator applications.

10.2.3 Magnesium diboride (MgB_2) wire

Magnesium diboride (MgB_2) is an inexpensive superconducting material. Although MgB_2 is not a new compound, it was only discovered to be a superconductor in 2001. MgB_2 has a critical temperature of 39 K. It is lower

Table 10.1 Performance and price comparison for some common superconductor materials.

Type	Price (€/m)	Current density (A/mm^2)	Flux density (T)	Temp. (K)
NbTi	0.4	10^3	5	4.2
Nb_3Sn	3	$1–4 \times 10^3$	5	4.2
MgB_2	4	10^2	3	20
Bi-2223	20	390	3	20
		10		50
YBCO	30	480	3	20
		190		50

Note: The flux density for YBCO, Bi-2223, is perpendicular to tape.
Source: Jensen *et al.*, 2012.

than the boiling point of nitrogen, but it is much higher than the critical temperatures of other alloys. For example, Niobium titanium (NbTi), which is commonly used in magnetic resonance imaging (MRI) devices and in particle accelerators, has a critical temperature of 9.2 K. MgB_2 can be cooled without using problematic and expensive liquid helium. Instead liquid hydrogen (boiling point 14 K) or liquid neon (boiling point 27 K) can be used for cooling. The current-carrying capacity of the MgB_2 wire improves significantly as the temperature is reduced below the critical temperature. For example, at 20 K a wire with 15–30% MgB_2 cross-section can carry 2000 A/mm^2 in a magnetic field of 1 T (Putti and Grasso, 2011).

The biggest advantage of MgB_2 is its simple structure. It is relatively easy and inexpensive to manufacture MgB_2 tapes and wires. It can be manufactured in lengths of a few kilometres. For example, in Razeti *et al.* (2008) an MRI device using 18 km of MgB_2 wire is presented. Furthermore, MgB_2 can be manufactured in various wire geometries and coil shapes, giving more freedom to coil design for machines.

10.3 Advantages of superconducting machines

A few MW sized superconducting machine prototypes have been tested in 1980s, but they couldn't be commercialized due to low operating temperatures and high operating cost. The discovery of high-temperature superconducting materials made it possible to build economically feasible applications. The superconducting machines have critical advantages over conventional machines. These advantages can be listed as follows.

10.3.1 High power density

Superconducting machines have higher power densities than conventional electrical machines. They operate with higher airgap flux densities due to

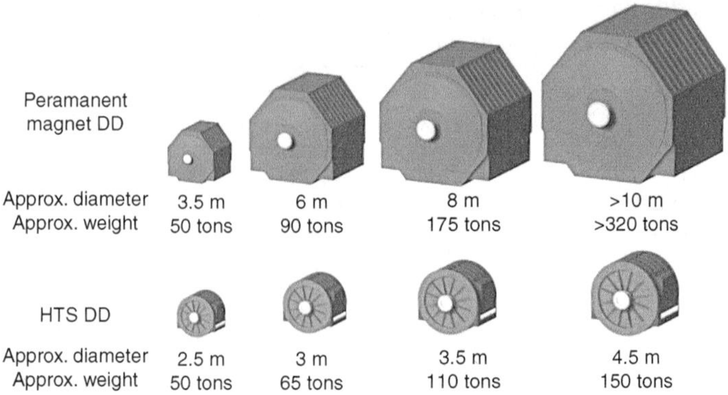

10.10 Size and mass comparison of direct drive permanent magnet generator and superconducting generator for different power ratings (AMSC, 2010).

higher magnetomotive force (MMF) created by superconducting wires that can conduct 100 A/mm^2 compared to 5–6 A/mm^2 of a copper coil.

In Fig. 10.10, direct drive permanent magnet generators (PMGs) are compared with superconducting generators in terms of mass and size. There are many studies in the literature that compare power densities of superconducting machines with conventional machines. Some of these are:

- Siemens 4 MVA superconducting generator has 82% of the volume of a similarly rated conventional generator and 63% of the weight (Klaus *et al.*, 2006).
- Half the losses of conventional motors in a package with half the volume can be achieved with superconducting machines (Aized *et al.*, 1999).
- An HTS ship propulsion motor made with BSCCO conductor, operating at 30 K, was one-third the size and half the weight of a conventional one, including the cooling system (Barnes *et al.*, 2005a).
- It is possible to reduce mass and volume by 40–50% with HTS machines (Klaus *et al.*, 2006).
- HTS machine applications can reduce motor losses by half and at the same time can reduce the overall machine weight and volume by half (Schiferl *et al.*, 2006).

10.3.2 Superior grid performance

In a slotless superconducting machine, the magnetic airgap is much larger than a conventional slotted machine. Large magnetic airgap reduces the synchronous reactance. A low synchronous reactance improves the dynamic stability of the machine during transient grid faults. Furthermore, the efficiency of the superconducting machines is almost independent of the power factor and field excitation current. Thus, superconducting machines can be over-excited for power factor correction. This eliminates the necessity of synchronous reactors and capacitor banks (Barnes *et al.*, 2005b).

10.3.3 Slow thermal ageing

Conventional electrical machine winding insulation materials degrade during thermal cycles of the machine. These thermal cycles reduce the lifetime of the insulation. In a superconducting machine, superconducting windings are always kept at cryogenic temperatures. The insulation in superconducting coils does not experience any thermal cycles. As a result, they have longer expected lifetime.

10.4 Challenges

Apart from the initial cost, the design of superconducting machines have many technical challenges. The most obvious one is having a cryocooler system to keep the superconducting windings cool enough. The rotor mechanical structure is another challenge that should be addressed properly.

10.4.1 Cooling

Cooling is the most challenging problem in the design of superconducting machines. Schiferl *et al.* (2006) state that the cost of cooling superconducting coils is roughly proportional to the inverse of the operating temperature in Kelvin. The relative cooling power required for various coolant gases is presented in Fig. 10.11. For a 4.5 MW superconducting machine operating at 30 K, the cooling power required is about 0.16% of the total power (0.5 m^3 in

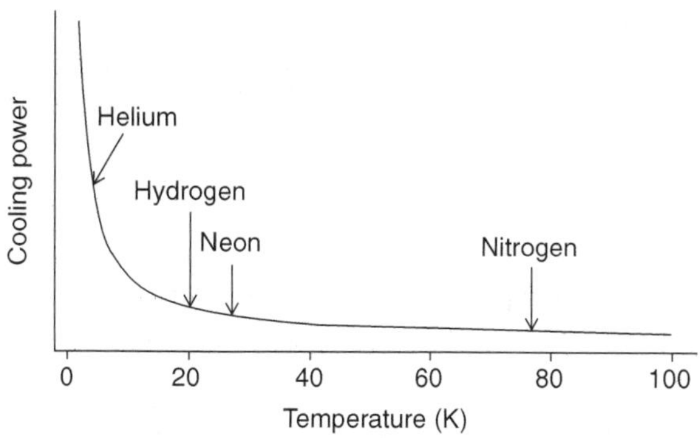

10.11 Comparison of cooling power required with different coolant gases.

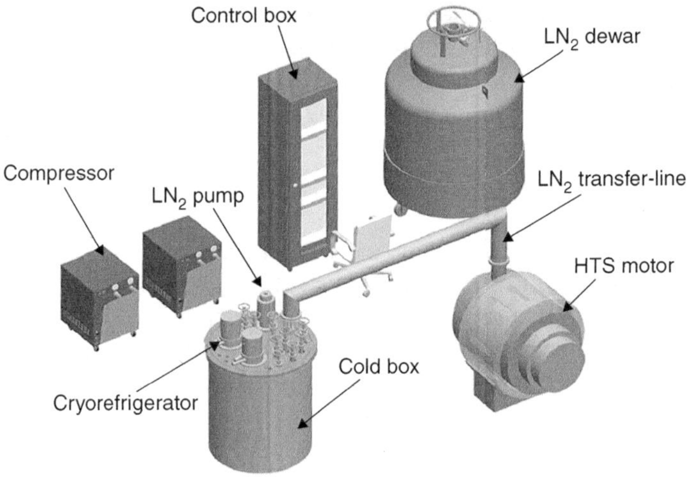

10.12 Layout of a typical liquid nitrogen cryocooler (Chen *et al.*, 2010).

volume). However, for a low-temperature superconducting machine operating at 4 K the cooling power required increases to 1.2% of its rated power, which has a significant effect on the overall efficiency (Schiferl *et al.*, 2006). This was the main technical and economic drawback with early superconducting machine applications.

After the discovery of HTS materials it became possible to use liquid nitrogen as a coolant, which is cheaper and more easily available than other cooling gases. The liquid nitrogen-based cryocooler technology is mature, reliable and has a maintenance cycle of 10 000 h (Schiferl *et al.*, 2006). A typical liquid nitrogen cooling system is presented in Fig. 10.12. However, the operating temperature is limited by the boiling point of nitrogen (77 K). If lower temperatures are required, neon, oxygen, hydrogen and helium are the other coolant options. Helium is very expensive and difficult to store; neon is expensive; hydrogen and oxygen are easy to obtain but they are hazardous and should be used with extensive care.

Decreasing the operating temperature of superconductors improves their performance, but it also increases the cooling power requirements. The operating temperature of the machine should be decided by considering the balance between the performance of a superconductor and the extra cost of the cooling system. The efficiency of the cooling system is about 1–10% depending on the operating temperature (Kautz, 2012). In other words, 100 W of electrical power is required to remove 1 W of heat loss.

Another challenge is the circulation of the coolant. Some cryocoolers use closed-circuit gas circulation (e.g. helium gas cryocoolers). In this case, an external pump is required for circulation. Another method is the thermosyphon method where the condensed gas is used. As the liquid evaporates it removes the heat and returns to the cryocooler. The biggest advantage of this method is that it does not need any mechanical assistance to circulate the coolant, so it is more reliable. The disadvantages of this method are listed by Tong (2010) as non-flexibility in the operating temperature, possible freezing of cold heads and non-homogeneous cooling of the rotor windings when the machine is stationary.

10.4.2 Rotor mechanical structure

The superconducting coils are placed in the rotor for most superconducting machine topologies. The challenge with that configuration is to keep the rotor at cryogenic temperatures and thermal losses to a minimum while the high torque from the shaft is being transferred. The rotor torque tube should therefore be a very good thermal insulator. Furthermore, high electrical and magnetic loading of superconducting machines results in high Lorentz forces acting directly on the superconducting wire, and the strain in the superconducting coils is limited by the material properties. A mechanical support should be carefully designed to prevent any damage to the superconducting wire. Carbon glass frames are usually used for rotor torque transmission, which increases the cost of the machine.

Another critical component for superconducting rotor is the cryocouplers that transfer cooling liquid from the stationary cooling system to the rotating rotor. A cryocoupler adds extra cost to the cooling system and decreases the overall reliability.

10.5 Superconducting machine topologies

Superconducting machine designs are limited by the AC characteristic of the superconducting wires described above. As a result, it is usual to employ superconductors only in the DC field winding of the machine. There are also some machine designs that use magnetized bulk-superconductor materials.

10.5.1 Conventional stator, superconducting rotor

The conventional stator, superconducting rotor is by far the most common configuration applied to many commercial products (Kalsi *et al.*, 2004; Barnes *et al.*, 2005b; Snitchler, 2010). In this configuration only the field winding in the rotor is superconducting. Conventional copper windings are used in the armature windings.

This means only the magnetic loading (i.e. flux density in the airgap) of the machine is improved compared to conventional machines, whereas the electrical loading (i.e. the current density in the armature winding) is the same as conventional machines.

The biggest challenge for this type of machine is the design of the superconducting rotor. The rotor requires a cryogenic coupler to transfer cooling gas. The long-term reliability of the cryogenic coupler is questionable. Electrical brushes or brushless exciters are also required to excite the field winding.

The machines in this category can be subdivided into four groups (Tong, 2010).

- Slotted stator with superconducting rotor with magnetic core (either warm or cold).
- Slotted stator with air-cored superconducting rotor.
- Slotless stator with magnetic cored superconducting rotor (either warm or cold).
- Slotless stator with air-cored superconducting rotor.

Rotor: magnetic or air cored

In conventional synchronous machines, the rotor has a magnetic core to minimize the reluctance of the flux path. The magnetic core can be eliminated in a superconducting machine. Extra MMF required for an air cored (non-magnetic) rotor can be easily supplied by the superconducting field winding.

An air cored rotor construction helps to reduce the mass of the machine. However, the air cored rotor requires more superconducting coil which increases cost. One advantage of the air cored rotor is the decrease in cold mass. Cooling power can be kept to a minimum and the machine can be cooled down to cryogenic temperatures in a much shorter time. It is also possible to use a magnetic cored rotor. Here there are two options: to cool down the magnetic core can with the superconducting coils – the cold rotor configuration – or to thermally isolate the magnetic core to reduce cold mass – a warm rotor configuration. The layouts of warm and cold iron rotor machines are presented in Fig. 10.13.

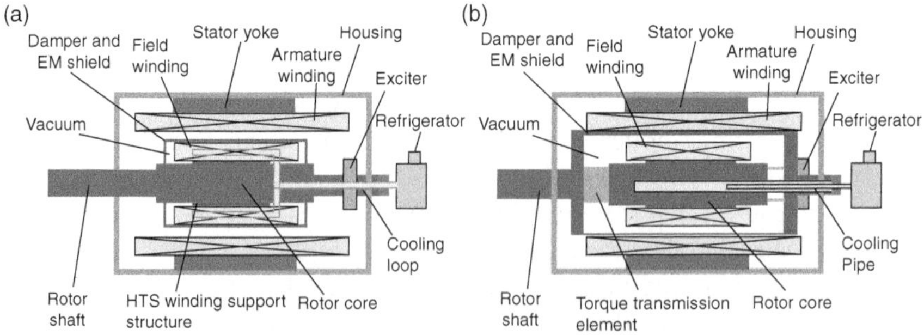

10.13 Magnetic rotor core arrangements (Klaus *et al.*, 2007). (a) Warm-iron rotor; (b) cold-iron rotor.

Stator: slotted or slotless

For the stator design, the first option is to use a slotted stator, which is no different from a conventional electrical machine. In this case, a superconducting rotor can be retro-fitted into an existing synchronous machine. For slotted machines, the power density of the machine is limited by the magnetic saturation in the stator teeth. The other option is to replace the magnetic stator teeth with a non-magnetic material such as carbon glass, as shown in Fig. 10.14. In this way, the higher flux densities can be obtained and the full advantage of using superconducting coils can be taken. Moreover, slot harmonics are eliminated and armature reaction is reduced, which increases the power quality of the machine. In slotless superconducting machines, due to high flux densities, Litz wires are preferred for armature winding in order to minimize the eddy current losses.

10.5.2 Homopolar superconducting machine

An air cored superconducting machine has higher power density but it is more expensive due to the excess amount of superconductor and it is more difficult to build. An alternative with reduced cost and increased robustness is homopolar superconducting synchronous machine (HSSM). The layout of a homopolar synchronous machine is given in Fig. 10.15. The biggest advantage of homopolar machines is the stationary superconducting field winding, which eliminates the cryocoupler and electrical exciter. Furthermore, the centrifugal and electromagnetic forces on the superconducting coil are eliminated by using a stationary superconducting coil. Homopolar machines are mechanically very robust, so they have been proposed for ship propulsion

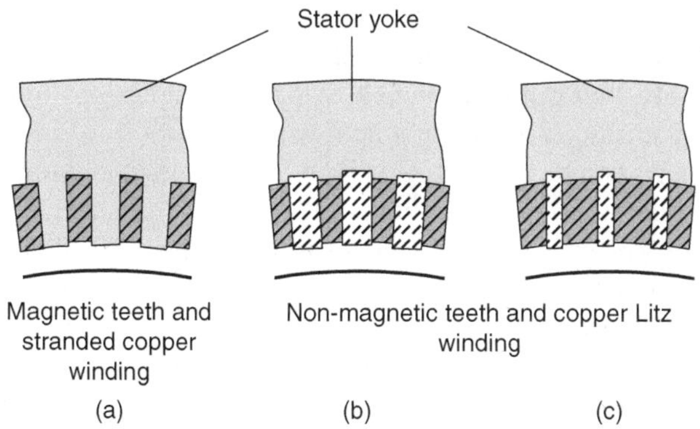

10.14 Different stator topologies of HTS machines. (a) Conventional stator; (b) high field stator; (c) high field, high ampere-turn loading stator (Klaus *et al.*, 2007).

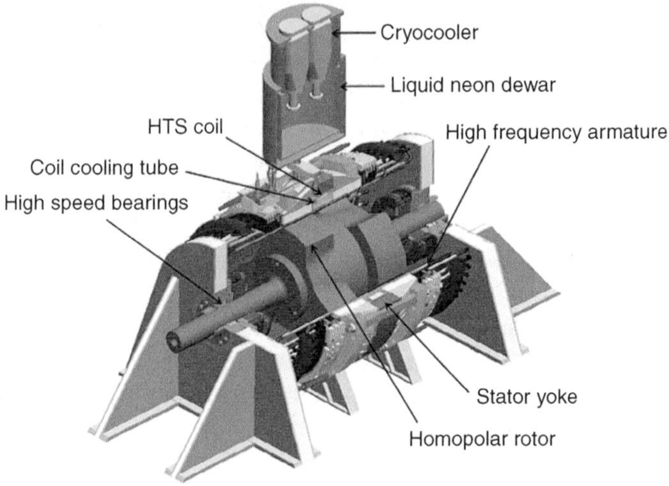

10.15 Layout of a homopolar superconducting machine (Zhang *et al.*, 2011).

motors. For example, in Lee *et al.* (2008), a conceptual design for a 5 MW, 230 rpm ship propulsion motor is presented. On the downside, homopolar superconducting machines are much heavier than the other types of superconducting machines due to large mass of the rotor core.

10.5.3 Axial-flux machines

Axial-flux superconducting machines are not very common and there are only a few prototypes with small power ratings. In Okazaki *et al.* (2006) and Matsuzaki *et al.* (2006), axial superconducting ship propulsion motor designed as presented in Fig. 10.16. Also in Masson *et al.* (2007a), an axial-flux HTS machine topology is proposed for an aircraft propulsion system. For a conventional stator (copper windings) axial-flux superconducting machine, there are two options for field winding: magnetized bulk superconducting field and superconducting field winding.

Superconductors can be magnetized by an external field and the field can be trapped in the superconductor due to non-decaying surface currents. Some bulk superconductors are capable of trapping magnetic flux densities up to 10 T. For a machine application, superconductors can be magnetized *in situ* or externally.

External magnetizing is not feasible for electrical machines since it requires the handling and assembly of highly magnetized superconductors at cryogenic temperatures. Moreover, if the cooling power is lost, all the superconductors demagnetize. In Matsuzaki *et al.* (2006), HTS bulk magnets are magnetized by the pulsed field magnetization method by using a pair of

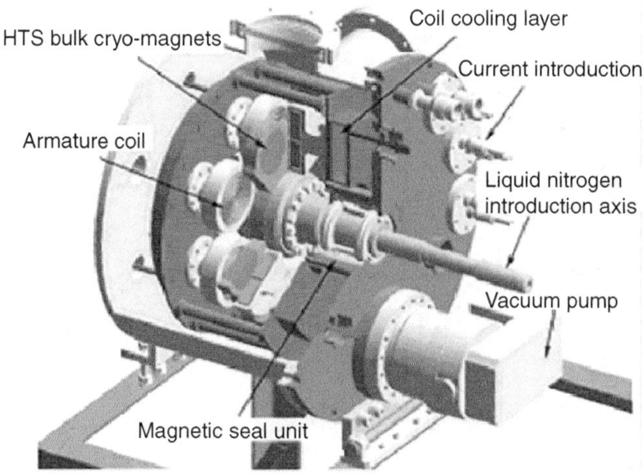

10.16 Schematic of a magnetized bulk HTS axial-flux machine (Matsuzaki *et al.*, 2006).

the armature coils. The disadvantage of this method is the limited flux densities that can be achieved by armature coils. The average flux density is not more than 0.5 T (Matsuzaki *et al.*, 2006). There are some novel flux pumping methods that magnetize superconductors gradually (see, e.g., Masson *et al.* (2007b) and Coombs *et al.* (2009)), but they have not yet been proved in larger applications.

10.5.4 All superconducting machines

The clear aim of superconducting machine designers is to replace copper windings with superconducting wire completely and to achieve a very high power dense machine with almost 100% efficiency. However, AC power loss in the superconducting coils is a serious obstacle. Barnes *et al.* (2005b) state that without an AC-tolerant HTS wire, the construction of a superconducting stator would not be possible or is limited to small power applications. In the University of Fukui (Sugimoto *et al.*, 2007) a promising prototype with superconducting armature and permanent magnet (PM) rotor has been built. The machine is designed for a ship propulsion system and has a rated output of 400 kW at 250 rpm.

10.6 Direct drive applications

Superconducting machines have higher torque densities than the conventional machines as presented in Section 10.3. This makes them a suitable option for direct drive applications. There are mainly two direct drive

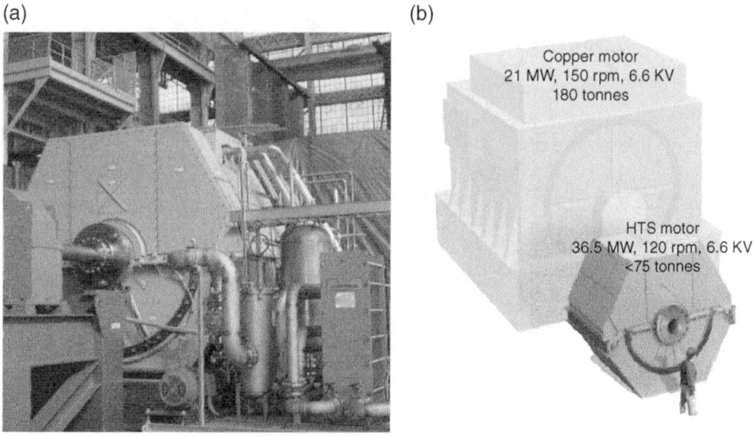

10.17 (a) AMSC's 36.5 MW superconducting ship propulsion motor; (b) comparison of superconducting machine with a conventional machine (Courtesy of AMSC).

10.18 Converteam 5 MW HTS ship propulsion motor coupled to test machine (Kalsi *et al.*, 2004).

applications of superconducting machines: ship propulsion systems and renewable energy power take-off systems.

10.6.1 Ship propulsion systems

The first direct drive superconducting machines have been applied in ship propulsion systems. Superconducting machines for ships promise increased reliability, decreased operating cost and increased cargo capacity. Lee *et al.* (2008) present a 5 MW, 230 rpm homopolar HTS machine. Another propulsion motor is the AMSCs 36.5 MW, 120 rpm superconducting machine with a rated torque of 2905 kNm, which is expected to be 75 tonnes (Kalsi *et al.*, 2004; Snitchler, 2010). The size of the machine is compared with a conventional machine in Fig. 10.17.

Converteam (recently acquired by GE) has previously been involved in manufacturing HTS ship propulsion systems. In Fig. 10.18 shows a 5 MW HTS machine coupled to 2.5 MW test machine. It can be clearly seen that the 2.5 MW conventional generator is the same size as the 5 MW superconducting machine.

10.7 Application to wind turbines

Installed renewable energy capacity has increased around the world over the last few decades and this trend is expected to continue for many years (Tong, 2010). The current cost of wind turbines is about €1.6 m per MW for onshore turbines and €3 m per MW for offshore wind turbines with a typical rating of €2.5 MW for onshore and €3.6 MW for offshore (Lesser and Müller, 2009). Lesser and Müller (2009) compare the costs of different generator types and suggest that the aim of the offshore wind turbine industry is to manufacture 10 MW turbines which cost €2.2 m per MW or less.

The average size for new offshore wind turbine installations is currently 5 MW. Larger wind turbines usually means reduced installation and maintenance cost per MWh. The onshore wind energy market is now quite mature and turbine size is limited by transportation issues and wind speeds. However, large offshore wind turbines can be assembled in port and transported in one piece. They also benefit from higher wind speeds and more constant wind regimes. According to market projections, annual world production by offshore wind turbines will increase from 2.7 to 6.2 GW by 2020 (Anon, 2009). Furthermore, the proportion of the turbines rated over 5 MW will increase significantly for the next decade (Fig. 10.19).

There are many studies that aim to build larger wind turbines. In a recently completed EU project (UpWind), the feasibility of 10 and 20 MW wind turbines was investigated, and one of the critical issues for very large wind turbines was stated to be excessive top-head mass (UpWind, 2011). The top-head mass is estimated to be 880 tonnes for a 20 MW wind turbine. If the tower-head mass is reduced, significant savings can be made in the structural, foundation and installation costs of the turbine. It is clear from the capacity of offshore installation vessels, currently around 300 tonnes (Lewis and Muller, 2007), that there is a demand for a high power density, direct drive generator topology.

One promising candidate to reduce the overall mass and volume of a large offshore wind turbine(up to 10 MW) power take-off system is superconducting generators (Kalsi *et al.*, 2004; Lesser and Müller, 2007, 2009). Lesser and Müller (2009) proposed that for generators larger than 6 MW, HTS generators are the only feasible solution. HTS machines can be scaled

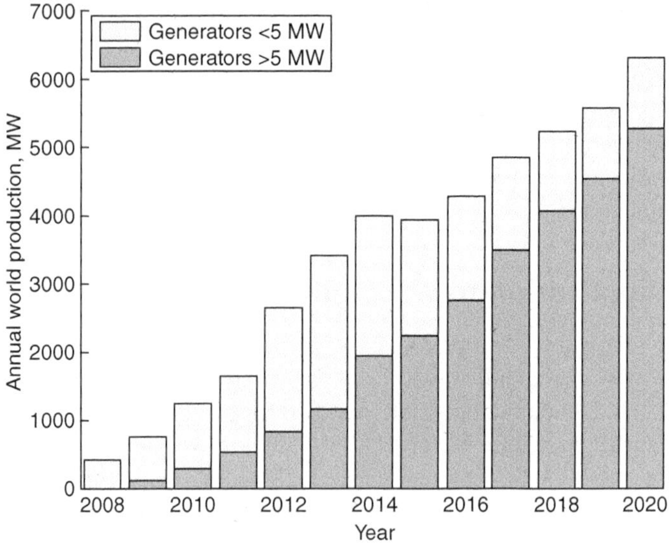

10.19 Emerging market projection for offshore wind energy converters (Data from Anon, 2009).

as the fifth power of mechanical dimensions in Kalsi *et al.* (2004) compared to third power (i.e. volume) for a conventional machine, which will become more dominant in direct drive machines. Lewis and Müller (2007) state that a 6 MW HTS direct drive generator may have 50% of the mass of a direct drive PM generator and the lower generator mass may allow the turbine to be transported and installed in one piece.

10.7.1 Mass comparison of direct drive systems

Data from different direct drive systems have been collected to compare mass-to-torque ratio of HTS machines with other types of generators. The result is presented in Fig. 10.20 and tabulated in Table 10.2. Note that some of the machines are final designs or commercial products (e.g. Enercon, Harakosan), whereas in other cases the designs are provisional or it is not clear how the structural mass is estimated. Although final designs tend to be heavier than initial designs, the graph should provide a good understanding of the torque density capability of HTS machines. The dashed line represents the ratio of generator mass to torque for permanent magnet machines, which is estimated as 25 kg/kNm (Bang *et al.*, 2008). The Enercon-E112, which has a high m/T ratio (66.6 kg/kNm) is the only copper field synchronous generator represented in the graph. The continuous line represents the linear trend line estimated using the HTS machines in the graph. The

Table 10.2 Torque density comparison of conventional and superconducting generators

	Manufacturer	Power (MW)	Speed (rpm)	Mass (t)	Torque (kNm)	Mass/T (kg/kNm)	Type
1	Harakosan (Duan and Harley, 2009)	1.5	18	47.2	796	59.3	PMG
2	The Switch (Duan and Harley, 2009)	3.8	21	81	1728	46.9	PMG
3	NewGen (Engstrom *et al.*, 2004)	4	19	36.4	2010	18.1	PMG
4	NREL-AMSC (Maples *et al.*, 2010)	3.1	12.5	90	2368	38	PMG[a]
5	Bang *et al.* (Bang *et al.*, 2009)	5	12	90.8	3979	22.8	TFPM
6		5	12	60.5	3979	15.2	TFPM[b]
7	NREL-AMSC (Maples *et al.*, 2010)	6	12.3	177	4658	38	PMG[c]
8		10	11.5	315	8304	37.9	PMG[d]
9	Bang *et. al.* (Bang *et al.*, 2008)	10	10	325	9543	34	PMG[e]
10	Enercon-E112 (Bang *et al.*, 2008)	4.5	13	220	3305	66.6	EESM
11	Lee *et al.* (Lee *et al.*, 2008)	5	230	44	207.6	212	HTSG[f]
12		5	230	18	207.6	86.7	HTSG[g]
13	Maki (Maki, 2008)	2	21.5	54	888	60.8	HTSG
14	NREL-AMSC (Maples *et al.*, 2010)	3.1	12.5	76	2368	32.1	HTSG
15	AMSC (Kalsi *et al.*, 2006)	36.5	120	75	2905	25.8	HTSG
16	Maki (Maki, 2008)	5	14.8	108	3226	33.5	HTSG
17	NREL-AMSC (Maples *et al.*, 2010)	6	12.3	97	4658	20.8	HTSG
18	Maki (Maki, 2008)	8	12	154	6366	24.2	HTSG
19	Converteam (Lewis and Muller, 2007)	8	12	100	6366	15.7	HTSG
20	NREL-AMSC (Maples *et al.*, 2010)	10	11.5	160	8303	19.3	HTSG
21	AMSC (Snitchler, 2010)	10	11	150	8681	17.3	HTSG
22	Abrahansen (Abrahamsen *et al.*, 2010)	10	10.4	88	9182	9.6	HTSG[h]

[a] The machine is not optimized for minimum mass.
[b] Ring-shaped transverse-flux permanent magnet machine (TFPM).
[c] The machine is not optimized for minimum mass.
[d] The diameter is limited to 4.3 m.
[e] Estimated mass.
[f] Homopolar superconducting machine.
[g] Air-cored HSSM.
[h] The machine is found to be economically infeasible.

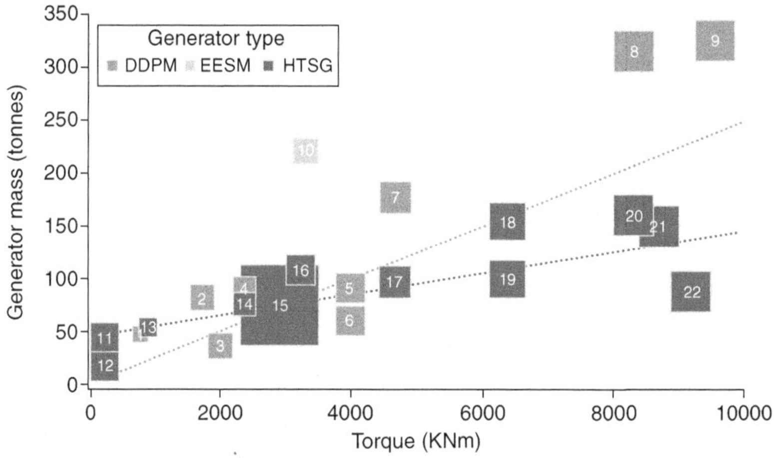

10.20 Mass of different large direct drive machines as a function of the torque, area of the square represents the power rating. EESM: Electrically excited synchronous machine; HTSG: High temperature superconducting generator; PMG: Permanent magnet generator. Dashed line represents generator mass-to-torque ratio (m/T) of 25 kg/kNm for permanent magnet machines as given by Bang *et al.* (2008). Continuous line is the linear trend line estimated using the HTS machines in the graph.

equation of the trend line is given in Equation [10.2]. It can be seen from the graph that HTS machines are generally lighter than PM generators for applications with torque requirements larger than 3000 kNm. There are two PMG topologies below the HTS machine trend line: a transversal flux permanent magnet machine (Bang *et al.*, 2009) and the NewGen topology which reduces the structural mass significantly (Engström *et al.*, 2004). It should be noted that the application of similar techniques in HTS machines can further decrease their mass.

$$\text{Mass } (t) = 0.011 \times \text{Torque (kN m)} + 45 \qquad [10.2]$$

10.7.2 Requirements for a superconducting wind turbine generator

In the first superconducting machine prototypes, high efficiency has been emphasized, but efficiency alone is insufficient. In order to penetrate the renewable energy market, high-temperature superconducting generator (HTSGs) have to prove that they are also more reliable than alternative power take-off systems such as direct-drive permanent magnet (DDPM) machines and geared induction machines (Abrahamsen *et al.*, 2010). Other requirements are availability and low cost.

Reliability

Offshore wind turbines operate in a harsher environment than those onshore. Furthermore, access to offshore wind turbines is much more difficult. Any repair is subject to weather and sea conditions, and in winter it may take weeks to access an offshore wind turbine (Tong, 2010). If a crane is required for a major replacement, it is also subject to the availability of appropriate vessels. This means weeks or months of lost energy income on top of the repair cost.

In a superconducting machine, the cryogenic system is the most critical component. Unfortunately, the cryocoolers cannot be eliminated until a room-temperature superconductor is discovered. However, the cooling system can be simplified and moving parts can be minimized by using a stationary superconducting coil design, which has many advantages (Gieras, 2008):

- No cryogenic coupler, more robust and cheap cooling system.
- No brushes or complex excitation systems.
- No centrifugal forces and transient torques that can damage the super-conducting material.
- Simpler winding support.

Availability

The availability of the generator directly affects the electricity generation income. The maintenance schedule should be decided at the design stage. The present generation of cryocoolers do require periodic maintenance. If the generator design requires the cryocooler to be turned off, then cooling times becomes a very important issue. For instance, a cold magnetic core is used by Wen *et al.* (2009) for a 100 kW generator with a cooling power of 120 W. The generator can be cooled down to operating temperature from ambient temperature in 18 h. Another example is Siemens' 4 MW machine, which has a total cooling power of 135 W. The cooling time for this machine is approximately 75 h (Klaus *et al.*, 2006). In order to increase availability, the cold mass of the generator should be minimized (a warm-iron core is preferable) and the cryocooler system should have some redundancy to allow formaintenance while keeping the superconducting coil at cryogenic temperatures.

Zenergy and Converteam have been involved in a research project on the development of maintenance-free cryocoolers (Lewis and Müller, 2007). Snitchler *et al.* (2010) described a particular refrigeration system which has an MBFR (mean time between failure rate) longer than 9 years. They also proposed using six to ten cryocoolers with $N+1$ redundancy with a maintenance cycle of 2 years for cryocoolers, fans and seals.

Cost

Maples *et al.* (2010) estimate the cost of an HTS machine to be lower than the cost of geared solutions for power ratings greater than 6.5 MW, and 8 MW for direct drive PM generators. This estimate is based on current HTS wire prices, but Lewis and Muller (2007) propose that the price of 2G HTS wires will beat the current copper prices as the demand for superconducting wire increases. The cost of 1G HTS wire decreased drastically in the mid-1990s (Tong, 2010). 2G wires first became commercially available after 2006 and within just a few years their price became comparable to that of 1G wire. Although 2G HTS wire cannot satisfactorily enter the market at current prices, performance is expected to continue improving with no significant increase in cost (Tong, 2010). These factors are taken into account by Lesser and Müller (2009) and it is estimated that the HTS machines will cost less than direct drive PM generators for applications larger than 5 MW.

The cost of superconducting generators can be calculated more accurately if the reduction in installation costs, increased availability and efficiency are also considered. In BVG (2012), the effect of superconducting generators on the overall cost of the project is estimated and it is stated that superconducting generators may reduce turbine nacelle capital expenditure by 2%. The high efficiency of superconducting generators will also result in a 2% increase in annual energy production (BVG, 2012).

10.7.3 Commercial projects

There are many commercial projects that aim to build a large direct drive superconducting generator for offshore wind turbines. One of the most ambitious is AMSC's SeaTitan project. AMSC, with experience of building a 36 MW superconducting motor, will manufacture a 10 MW, 10 rpm offshore wind turbine. SeaTitan is planned to be installed on a turbine with a diameter of 190 m. The generator will have 96% efficiency at rated load and is estimated to be 150–80 tonnes with a 4.5–5 m diameter (Snitchler, 2010). The overall weight of the nacelle will be less than 500 tonnes (AMSC, 2010).

SeaTitan uses YBCO-based Amperium wires, and the operating temperature of the generator is about 30–40 K (AMSC, 2010). The cooling system uses compressed helium. To increase the reliability of the turbine, redundant cryogenic cooling systems will be installed. Thus, even if one of the cryogenic coolers fails, the rest can continue to operate without affecting power generation. Long-life seals are preferred to increase the mean time between maintenance. Most of the subcomponents (such as cryogenic cooler compressors, power converters) are placed at the bottom of tower for easy access and replacement (Fischer, 2011). The cooling system in the rotor is presented in Fig. 10.21. One interesting feature of the rotor is that the cold

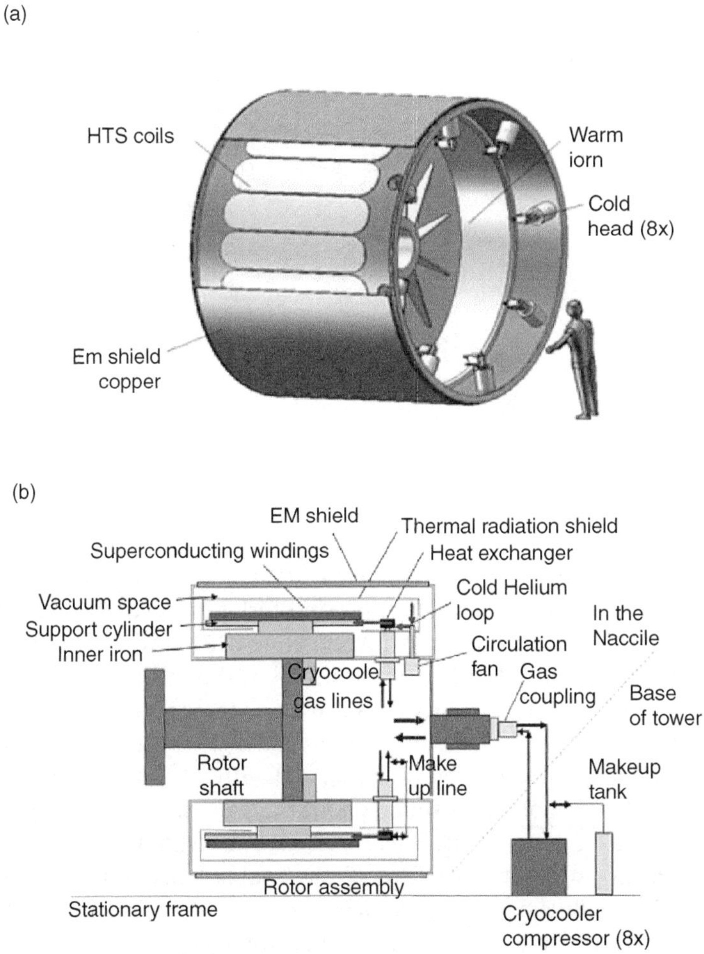

10.21 AMSC's 10 MW SeaTitan generator (Courtesy of AMSC). (a) Superconducting rotor; (b) layout of cooling system (AMSC, 2010).

heads (eight in total) are placed in the rotor to minimize cooling loss and increase reliability.

A slotless stator configuration is used to eliminate saturation in the stator teeth. The stator windings are water cooled. The mechanical airgap is 20 mm and the magnetic airgap is significantly larger due to its air cored structure (AMSC, 2010). To increase its wind turbine manufacturing capabilities, AMSC acquired a 25% stake in a blade manufacturer (Blade Dynamics) (AMSC, 2012).

They also planned to acquire a DDPM generator manufacturer (The Switch), but the agreement ran into finance problems (ams, b). AMSC expects SeaTitan to penetrate the market by 2015 (Fischer, 2010). Advanced Magnet

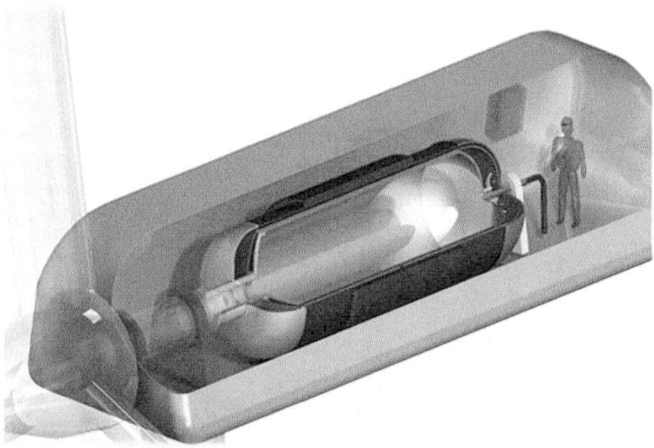

10.22 AML's 10 MW 10 rpm fully superconducting generator (AML, 2012).

Lab (AML) is one of the six companies that were awarded $7.5 million in total to develop the next generation wind turbines. AML and its subsidiary company AML Energy are now investigating a 10 MW, 10 rpm fully superconducting generator. MgB_2 wires are used in the machine and the operating temperature is 20 K. Helium gas is used as the coolant. The machine has a length of 5 m and a diameter of 2 m (AML, 2012). The weight including the cooling system will be 50 tonnes (AML, 2012). The generator is shown in Fig. 10.22. A double-helix superconducting windings technology is used to minimize the AC loses on the armature coils and to limit the flux penetrating into the coils. Another advantage of the double-helix configuration is perfectly sinusoidal voltage output, which eliminates any harmonics and reduces vibration. AMLplan to build a full-scale prototype by 2015 (AML, 2012).

Converteam, now General Electric Power Conversion, completed a feasibility study with Zenergy in 2006 on direct drive HTS generators for wind turbines. Following this a project to manufacture a full-scale superconducting wind turbine was started (supported by a grant from the UK government) (Tong, 2010). The conceptual design is rated at 8 MW, 12 rpm, for a 160 m diameter turbine (Lewis and Muller, 2007). The design has a diameter of 5 m and axial length of 2.2 m, and weighs approximately 100 tonnes.

Converteam also installed a 1.7 MW, 214 rpm superconducting generator at a hydroelectric power station in Germany (Fair *et al.*, 2010). The superconducting rotor is retro-fitted to a conventional synchronous generator. Bi-2223 superconducting tapes are used in the generator (Fig. 10.23). The operating temperature is 30 K (Fair *et al.*, 2010).

General Electric (GE) is continuing to research superconducting generators with their clean technology brand 'Ecomagination'. Like AML, GE was

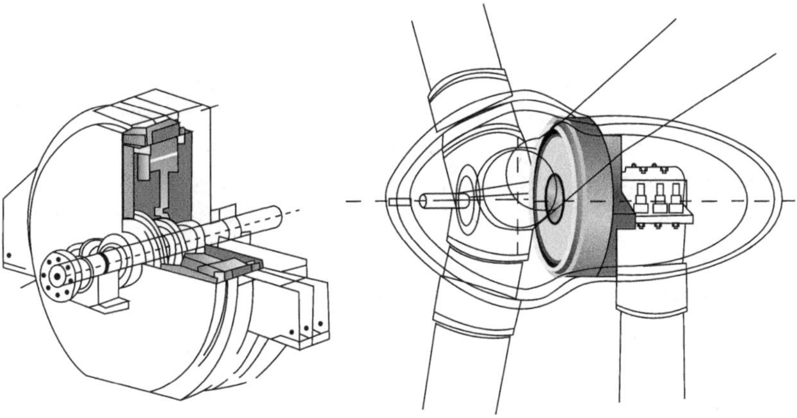

10.23 Converteam 1.7 MW 214 rpm HTS hydro generator and 8 MW 12 rpm direct drive HTS wind turbine generator (Courtesy of GE).

awarded a grant by the US Department of Energy for next-generation wind turbines. They are designing a 10–15 MW low-temperature superconducting (LTS) generator. The generator has stationary superconducting field winding and rotating armature. At 4 K, the operating temperature will be lower than any other designs (Jensen *et al.*, 2012). The generator may NbTi wires, which are extensively used in MRI devices. The superconductor cost can be reduced with NbTi wires, but the cooling power required to operate the machine at 4 K is higher (Jensen *et al.*, 2012).

There are also some conceptual designs for superconducting wind turbines. Terao *et al.* (2012) propose a 10 MW, 10 rpm fully superconducting machine. HTS tapes are used in rotor for DC field winding and MgB_2 wires for the armature winding. The operating temperature of the generator is 20 K. The diameter of the generator is 3.7 m and it weighs around 35 tonnes (excluding structural mass). The AC loss in the MgB_2 wires is estimated as 1 kW. The biggest drawback of the generator is the excess amount of superconducting wire. It requires 250–300 km each of HTS tape and MgB_2 wire, which makes the design unfeasible under current manufacturing capabilities.

Quddes *et al.* (2011) present another 10 MW, 10 rpm superconducting machine. The machine uses a novel transverse-flux configuration. The maximum flux density in the airgap is more than 4 T. The superconducting coils are circular and hence easy to manufacture. On the other hand, the mechanical stress on the coils is very large and the support structure has to be stiff enough to cope with these forces.

Keysan and Mueller (2011) present an axial-flux superconducting generator topology. The generator has a stationary superconducting field winding and stationary armature winding. The design eliminates the cryocouplers and electrical brushes, and it is expected to have high reliability. However, the iron-cored rotor limits the maximum flux density and increases the generator mass.

Although there is growing interest in direct drive generators, it may take a long time to prove the concepts and penetrate the market. It is estimated that in 2014 the first commercial prototypes will enter the wind turbine market, continuing to penetrate it until 2017, but government support is necessary to turn these prototypes into successful commercial products (Anon, 2006).

10.8 Application to wave energy

The wind turbine industry is quite mature and benefits from standardized components and commercialization, but wave energy converters (WECs) are immature and there is no standard power take-off system. There are some WECs that use hydraulic power take-off systems such as Pelamis Wave Power and Aquamarine Power (Alldritt and Hopwood, 2010). The energy conversion process in a hydraulic system involves many intermediate stages (i.e. using hydraulic pumps, high pressure pipes, hydraulic motors and electrical generators). Each of these components has the potential to cause some unscheduled maintenance. The maintenance for marine energy converters is difficult and costly. There are some direct drive PMGs for point observer type WECs (Leijon *et al.*, 2006; Mueller *et al.*, 2007). Direct drive generators are proposed in order to increase the overall reliability of the system.

Superconducting machines can also be an alternative to direct drive PM machines. The high current-carrying capability of superconducting materials makes it possible to create very large forces that are required for low-speed direct driveWECs. Superconducting linear machines are usually used for electromagnetic launcher applications (Stumberger *et al.*, 2004) and magnetic levitation trains (Sawada, 2009). In Jin *et al.* (2011), a self-levitated HTS linear machine with bulk superconducting magnets is proposed for this kind of application. HTS magnets, which can be much stronger than rare-earth permanent magnets, are an efficient way to create an excitation field in the machine, but the superconducting magnets are demagnetized if a fault occurs in the cooling system. Without amagnetizing system *in situ*, they are not suitable for WECs because of reliability issues. Alternatively, superconducting wires can be used to excite the linear machine (Yen *et al.*, 2010). In Terao *et al.* (2010), a double-sided air cored linear synchronous motor that uses both high-temperature superconducting coils and bulks is presented. The average thrust density of this machine is 47.3 kN/m^2, which is much higher than that of a PM linear machine. Also, in Keysan and Mueller (2012) an air-cored superconducting linear machine suitable for low-speed direct drive applications is presented. The machine is similar to the rotational machine presented by authors in Keysan and Mueller (2011) for a direct drive wind turbine. The machine is shown in Fig. 10.24. The thrust

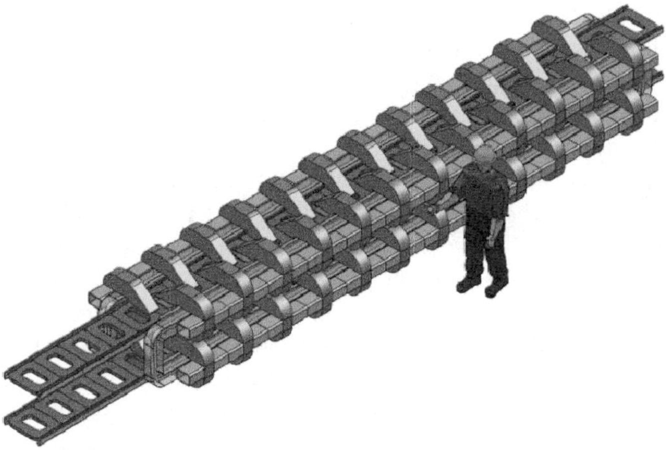

10.24 Linear superconducting generator for direct drive WECs. Power rating is 640 kW, 2.2 m/s (Keysan and Mueller, 2012).

density of the machine is 30.3 kN/m^2, less than that in Terao *et al.* (2010) due to the iron-cored translator.

10.9 Conclusion

Superconducting electrical machines have great potential in terms of power density and efficiency. It is possible to reduce the mass and volume by half with superconducting machines. This becomes more dominant at high power ratings. This chapter has shown that superconducting machines become lighter than direct drive PMGs for applications that demand more torque than 3000 kNm (e.g. a 5 MW, 13 rpm wind turbine). For wind turbines lighter generator means reduced nacelle mass, and therefore reduced foundation and installation costs.

The current price of the superconducting wires is one of the biggest obstacles to the commercialization of superconducting electrical machines. The price of 2G wires is coming down and it is estimated that in a decade 2G wires will be cheaper than copper wires with same current rating. This assumes that the production volume will increase significantly in the following years. It is clear that superconducting machines will increase demand for superconducting tapes and help to reduce prices. However, until that time, the initial prototypes need to be supported by governments and research funding for a successful entry to the market.

The reliability of the cooling systems is another challenge. High-temperature superconducting materials allowed the use of liquid nitrogen as a coolant, which is cheap and abundant. Liquid nitrogen cryocoolers are used in many industries and they have become more reliable in recent years.

Nevertheless, cryogenics introduce extra moving parts, pipes and seals that need to be maintained regularly. Furthermore, there are some applications that prefer to use liquid neon or hydrogen, which is more problematic, to reduce the operating temperature in order to maximize the performance of the superconducting wires.

The reliability of the cryocoolers has been proved in controlled environments, such as in a hospital for an MRI device. However, reliability inside the nacelle of an offshore wind turbine is still questionable. There are novel methods that aim to increase the system reliability under these conditions. AMSC's 10 MW, 10 rpm SeaTitan uses redundant cooling system with long-life seals. Furthermore, most of the components are placed at the bottom of the tower for easy access and replacement (AMSC, 2010; Fischer, 2011). AML uses a double-helix winding configuration to reduce AC losses and vibration (AML, 2012). General Electric preferred a stationary superconducting winding in order to eliminate couplers and brushes.

All these new technologies should be tested in real operating conditions. It may take a decade to prove that superconducting generators are at least as reliable as conventional generators. It should also be shown that superconducting generators are more cost-effective in the long run with reduced installation costs and increased efficiency. After that, the wind industry market can adapt the superconducting generators, which may take another decade. The development of WECs is still at a very early stage. There are only a few conceptual designs and very small prototypes. Commercial superconducting direct drive linear machines will be only possible after a success in the wind energy market.

10.10 References

Abrahamsen, A. B., Mijatovic, N., Seiler, E., Zirngibl, T., Træholt, C., Nørgard, P. B., Pedersen, N. F., Andersen, N. H. and Østergard, J. (2010). Superconducting wind turbine generators. *Superconductor Science and Technology*, **23**(3):034019.

Aized, D., Gamble, B., Sidi-Yekhlef, A., Voccio, J., Driscoll, D., Shoykhet, B. and Zhang, B. (1999). Status of the 1000 HP HTS motor development. IEEE Transactions on applied superconductivity, **l9**–2:1197–1200.

Alldritt, D. and Hopwood, D. (2010). Renewable energy in Scotland. *Renewable Energy Focus*, **11**(3):28–33.

AML, Advanced Magnet Lab (2012). Web page. Available from: http://www.aml-cleanenergy.com/ [Accessed 8 August 2012].

AMSC web page (2012), AMSC Acquires 25% Stake in Blade Dynamics Ltd. Available at: http://ir.amsc.com/releasedetail.cfm?ReleaseID=642671 [Accessed 13 July 2012].

AMSC pulls out of The Switch acquisition. Available at: htp://www.renewableenergyfocus.com/ view/21755/amsc-pulls-out-of-the-switch-acquisition/ [Accessed 13 July 2012].

AMSC (2010). Concepts for High Power Wind Turbines – Introducing HTS Technology. World Green Energy Forum.

Anon. (2006). *High Temperature Superconductivity Market Readiness Review – Navigant Consulting*. Technical report, Department of Energy, USA.

Anon. (2009). Global Offshore Wind Energy Markets and Strategies: 2009–2020, Emerging Energy Research. Technical report.

BVG Associates (2012). Offshore wind cost reduction path ways. Technical Report Available: http://www.thecrownestate.co.uk/media/305086/BVG%20OWCRP%20technology%20work%20stream.pdf

Bang, D., Polinder, H., Shrestha, G. and Ferreira, J. A. (2008). Review of Generator Systems for Direct-Drive Wind Turbines. In European Wind Energy Conference & Exhibition, Belgium, pp. 1–11.

Bang, D.-j., Polinder, H., Shrestha, G. and Ferreira, J. A. (2009). Ring-shaped transverse flux PM generator for large direct-drive wind turbines. In 2009 International Conference on Power Electronics and Drive Systems (PEDS), pp. 61–6.

Barnes, P., Rhoads, G., Tolliver, J., Sumption, M. and Schmaeman, K. (2005a). Compact, lightweight, superconducting power generators. *IEEE Transactions on Magnetics*, **41**(1):268–73.

Barnes, P., Sumption, M. and Rhoads, G. (2005b). Review of high power density superconducting generators: Present state and prospects for incorporating YBCO windings. *Cryogenics*, **45**(10–11):670–86.

Chen, A., Liu, X., Xu, F., Cao, J. and Li, L. (2010). Design of the cryogenic system for a 400 kW Experimental HTS Synchronous Motor. *IEEE Transactions on Applied Superconductivity*, pp. 3–6.

Coombs, T. A., Hong, Z., Yang, Y. and Rawlings, C. D. (2009). The next generation of superconducting permanent magnets: The flux pumping method. *IEEE Transactions on Applied Superconductivity*, **19**(3):2169–73.

Duan, Y. and Harley, R. G. (2009). Present and future trends in wind turbine generator designs. 2009 IEEE Power Electronics and Machines in Wind Applications, pp. 1–6.

Engstrom, S., Hernnas, B., Parkegren, C. and Waer-nulf, S. (2004). Development of NewGen – a new type of direct-drive generator. In European Wind Energy Conference-EWEC, volume 4, pp. 22–5.

Fair, R., Lewis, C., Eugene, J. and Ingles, M. (2010). Development of an HTS hydroelectric power generator for the hirschaid power station. *Journal of Physics: Conference Series*, **234**(3):032008.

Fischer, M. (2010). Superconductors on the high Seas. *Wind Systems*, pp. 76–84.

Fischer, M. (2011). High temperature superconductors enable the world's largest and most powerful wind turbine. *Wind Power*, pp. 60–2.

Gieras, J. F. (2008). *Advancements in Electric Machines (Power Systems).* Springer.

Gurevich, A. (2011). To use or not to use cool superconductors? *Nature materials*, **10**(4):255–9.

Jensen, B. B., Mijatovic, N. and Abrahamsen, A. B. (2012). Advantages and challenges of superconducting wind turbine generators. In International Conference E/E Systems for Wind Turbines, number May, pp. 1–29, Bremen.

Jin, J. X., Zheng, L. H., Guo, Y. G. and Zhu, J. G. (2011). Performance characteristics of an HTS linear synchronous motor with HTS bulk magnet secondary. *IEEE Transactions on Industry Applications*, **47**(6):2469–77.

Kakani, S. (2009). *Superconductivity*. Anshan Ltd.

Kalsi, S., Gamble, B., Snitchler, G. and Ige, S. (2006). The status of HTS ship propulsion motor developments. IEEE Power Engineering Society General Meeting, 5pp.

Kalsi, S., Weeber, K., Takesue, H., Lewis, C., Neumueller, H. and Blaugher, R. (2004). Development status of rotating machines employing superconducting field windings. *Proceedings of the IEEE*, **92**(10):1688–704.

Kalsi, S. S. (2011). *Applications of High Temperature Superconductors to Electric Power Equipment*. John Wiley & Sons, Inc., Hoboken, NJ, USA.

Kautz, R. (2012). 100 years of superconductivity. *Physics Today*, **65**(7):54.

Keysan, O. and Mueller, M. A. (2011). A homopolar HTSG topology for large direct-drive wind turbines. *IEEE Transactions on Applied Superconductivity*, **21**(5):3523–31.

Keysan, O. and Mueller, M. A. (2012). A linear superconducting generator for wave energy converters. In Power Electronics, Machines and Drives, PEMD, IET Conference on, p. B134, Bristol.

Klaus, G., Nick, W., Neumueller, H., Nerowski, G. and McCown, W. (2006). Development of high-temperature superconducting electrical machines at Siemens AG. In International Conference on Electrical Machines, ICEM, pp. 1–6.

Klaus, G., Wilke, M., Frauenhofer, J., Nick, W. and Neumueller, H. (2007). Design challenges and benefits of HTS synchronous machines. IEEE Power Engineering Society General Meeting, pp. 1–8.

Lee, S.-H., Hong, J., Kwon, Y., Jo, Y.-S. and Baik, S. K. (2008). Study on homopolar superconductivity synchronous motors for ship propulsion applications. *IEEE Transactions on Applied Superconductivity*, **18**(2):717–20.

Leijon, M., Danielsson, O., Eriksson, M., Thorburn, K., Bernhoff, H., Isberg, J., Sundberg, J., Ivanova, I., Sjostedt, E., Agren, O. and others (2006). An electrical approach to wave energy conversion. *Renewable Energy*, **31**(9):1309–19.

Lesser, M. and Müller, J. (2009). *Superconductor Technology Generating the Future of Offshore Wind Power*. Renewable Energy World Conference, Cologne, Germany, pp. 1–10.

Lewis, C. and Muller, J. (2007). A Direct Drive Wind Turbine HTS Generator. 2007 IEEE Power Engineering Society General Meeting, pp. 1–8.

Maki, N. (2008). Design study of high-temperature superconducting generators for wind power systems. *Journal of Physics: Conference Series*, (**97**)012155.

Maples, B., Hand, M. and Musial, W. (2010). *Comparative Assessment of Direct Drive High Temperature Superconducting Generators in Multi-Megawatt Class Wind Turbines*. Technical Report October, National Renewable Energy Laboratory (NREL), Golden, CO.

Masson, P. J., Breschi, M., Tixador, P. and Luongo, C. A. (2007a). Design of HTS Axial Flux Motor for Aircraft Propulsion. *IEEE Transactions on Applied Superconductivity*, **17**(2):1533–6.

Masson, P. J., Pienkos, J. E. and Luongo, C. A. (2007b). Scaling up of HTS motor based on trapped flux and flux concentration for large aircraft propulsion. *IEEE Transactions on Applied Superconductivity*, **17**(2):1579–82.

Matsuzaki, H., Kimura, Y., Ohtani, L., Marita, E., Ogata, H., Izumi, M., Ida, T., Sugimoto, H., Miki, M., Kitano, M. and Fujimoto, H. (2006). Mechanical and cryogenic design of a synchronous rotating machines with HTS pole-field magnets. IEEE Power Engineering Society General Meeting, p. 4.

Mueller, M., Polinder, H. and Baker, N. (2007). Current and novel electrical generator technology for wave energy converters. 2007 IEEE International Electric Machines & Drives Conference, (1):1401–6.

Ohsaki, H., Sekino, M., Suzuki, T. and Terao, Y. (2009). Design study of wind turbine generators using superconducting coils and bulks. 2009 International Conference on Clean Electrical Power, pp. 479–84.

Okazaki, T., Sugimoto, H. and Takeda, T. (2006). Liquid nitrogen cooled HTS motor for ship propulsion. In IEEE Power Engineering Society General Meeting, p. 6.

Putti, M. and Grasso, G. (2011). MgB2, a two-gap superconductor for practical applications. *MRS Bulletin*, **36**(08):608–13.

Quddes, M. R., Sekino, M., Ohsaki, H., Kashima, N. and Nagaya, S. (2011). Electromagnetic design study of transverse flux enhanced type superconducting wind turbine generators. *IEEE Transactions on Applied Superconductivity*, **21**(3):1101–4.

Razeti, M., Angius, S., Bertora, L., Damiani, D., Marabotto, R., Modica, M., Nardelli, D., Perrella, M. and Tassisto, M. (2008). Construction and operation of cryogen free MgB2 magnets for open MRI systems. *IEEE Transactions on Applied Superconductivity*, **18**(2):882–6.

Sawada, K. (2009). Outlook of the superconducting Maglev. *Proceedings of the IEEE*, **97**(11):1881–5.

Schiferl, R., Flory, A., Umans, S. and Livoti, W. (2006). High temperature superconducting synchronous motors: Economic issues for industrial applications. IEEE Industry Applications Society 53rd Annual Petroleum and Chemical Industry Conference, pp. 1–9.

Snitchler, G. (2010). Progress on high temperature superconductor propulsion motors and direct drive wind generators. Power Electronics Conference (IPEC), 2010, (c):5–10.

Stumberger, G., Aydemir, M., Zarko, D. and Lipo, T. (2004). Design of a linear bulk superconductor magnet synchronous motor for electromagnetic aircraft launch systems. *IEEE Transactions on Appiled Superconductivity*, **14**(1):54–62.

Sugimoto, H., Tsuda, T., Morishita, T., Hondou, Y., Takeda, T., Togawa, H., Oota, T., Ohmatsu, K. and Yoshida, S. (2007). Development of an axial flux type PM synchronous motor with the liquid nitrogen cooled HTS armature windings. *IEEE Transactions on Applied Superconductivity*, **17**(2):1637–40.

SuperOx Web Page. Available at: http://www.superox.ru/ [Accessed 13 August 2012].

Terao, Y., Sekino, M. and Ohsaki, H. (2010). Design study of linear synchronous motors using superconducting coils and bulks. In Power Electronics Conference (IPEC), 2010 International, pp. 1760–5. IEEE.

Terao, Y., Sekino, M. and Ohsaki, H. (2012). Electromagnetic design of 10 MW class fully superconducting wind turbine generators. *IEEE Transactions on Applied Superconductivity*, **22**(3):5201904

Tong, W. (2010). *Wind Power Generation and Wind Turbine Design*. WIT Press.

UpWind (2011). UpWind – Design Limits and Solutions for Very Large Wind Turbines – EU 6th Frame Project. Technical report.

Wen, H., Bailey, W., Goddard, K., Al-Mosawi, M. K., Beduz, C. and Yang, Y. (2009). Performance test of a 100 kW HTS generator operating at 67 K-77 K. *IEEE Transactions on Applied Superconductivity*, **19**(3):1652–5.

Yen, F., Li, J., Zheng, S. J., Liu, L., Ma, G. T., Wang, J. S. and Wang, S. Y. (2010). A single-sided linear synchronous motor with a high temperature superconducting coil as the excitation system. *Superconductor Science and Technology*, **23**(10):105015.

Zhang, T., Haran, K., Laskaris, E. and Bray, J. (2011). Design and test of a simplified and reliable cryogenic system for high speed superconducting generator applications. *Cryogenics*, **51**(7):380–3.

Index